Augustus Jay Du Bois

Tables for Finding the Strains in Railway Bridge Trusses

Under the Action of a Concentrated Load System

Augustus Jay Du Bois

Tables for Finding the Strains in Railway Bridge Trusses Under the Action of a Concentrated Load System

ISBN/EAN: 9783744689946

Printed in Europe, USA, Canada, Australia, Japan

Cover: Foto ©berggeist007 / pixelio.de

More available books at **www.hansebooks.com**

TABLES

FOR

FINDING THE STRAINS

IN

RAILWAY BRIDGE TRUSSES

UNDER THE ACTION OF A

CONCENTRATED LOAD SYSTEM.

REPRINTED FROM STRAINS IN FRAMED STRUCTURES.

SECOND EDITION.

BY

A. J. DUBOIS,

PROFESSOR OF CIVIL ENGINEERING IN SHEFFIELD SCIENTIFIC SCHOOL OF YALE COLLEGE.

PREFACE.

THESE Tables are intended to aid those engaged in the calculation of trusses. The system of concentrated loads annexed, is very closely that generally laid down in the specifications of our principal railroads. The use of the tables will be found sufficiently illustrated for all those for whom they are designed. The principles upon which the calculations are based are given at length in the author's work, "*Strains in Framed Structures*," from which these Tables are reprinted.

December 10, 1884.

TABLE I.

CONCENTRATED LOAD SYSTEM.

SHEAR.

TABLE I.

CONCENTRATED LOAD SYSTEM.
SHEAR.

EXPLANATION OF TABLE.

THE load-system adopted is given in Table I. It consists of two locomotives with tenders, followed by train; the wheel-weights and distances being given in the first column. The locomotive agrees closely with the "standard consolidated engine," usually specified by the best railroad and bridge companies. It is believed that the load-system assumed is such as will give, in every case, results slightly in excess of the heaviest possible loading which can actually occur in practice.

The use of Table I. is quite simple. The length of span is l. The load always comes on *from right*. The maximum shear at any point, distant x from *the left end*, is given by one of *four* formulæ. These formulæ, and the limits between which they hold good, are given at the head of the table. For the *first eight feet* of any span we must use formula (I.). From $x = 8$ up to a certain *limit* we must use (II.). This limit is given in the table to right, for every length of span from 27 to 300. Thus, for a span of 100 feet, this limit is 47½ feet *from the left end*. From this limit, up to $x = l - 8$, we must use formula (III.). At the limit, both (II.) and (III.) give the same result. For the *last eight feet* of span we use formula (IV.).

Thus, for instance, for a span of 200 feet, we must use (I.) from $x = 0$ up to $x = 8$. From $x = 8$ up to $x = 89\frac{1}{2}$ we must use (II.). From $x = 89\frac{1}{2}$ up to $x = 192$ we must use (III.). From $x = 192$ up to $x = 200$ we must use (IV.).

The table gives the values of P and A for the corresponding value of $l - x$, or $l - x - 8$.

Thus, suppose we wish the maximum shear for a span of 200 feet, at 7, 80, 100, and 195 feet from end, load coming on always from right.

For $x = 7$ we use formula (I.). We have, then, $l = 200$, $l - x = 193$. From table, we see that for any value of $l - x$ between 190 and 195 we have $P = 527000$, and $A = 41730000$. Hence, for $x = 7$,

$$\text{Shear} = \frac{527000}{200}193 - \frac{41730000}{200} = 299905 \text{ lbs.}$$

In like manner, for $x = 80$, we have $l - x = 120$, $l - x + 8 = 128$; and, since the limit is 89½, we must use formula (II.). We have from table, for $l - x = 120$, $P = 383000$, and $A = 19410000$; and, hence,

$$\text{Shear} = \frac{15000}{200}128 + \frac{383000}{200}120 - \frac{19410000}{200} - 15000 = 127350 \text{ lbs.}$$

Observe that the table gives these values of P and A for any value of $l - x$ *between 115 and 120, including* 120. For any value of $l - x$ between 120 and 125, *including* 120, we have, also, $P = 395000$, $A = 20850000$; and these values in formula (II.) would give the same result. Thus,

$$\text{Shear} = \frac{15000}{200}128 + \frac{395000}{200}120 - \frac{20850000}{200} - 15000 = 127350 \text{ lbs.}$$

For $x = 100$, we have $l - x = 100$, $l - x - 8 = 92$: and, since the limit is 89½, we must use formula (III.). We have from table, for $l - x - 8 = 92$, $P = 335000$, $A = 14370000$; and, hence,

$$\text{Shear} = \frac{15000}{200}100 + \frac{335000}{200}92 - \frac{14370000}{200} = 89750 \text{ lbs.}$$

For $x = 195$, since this falls within the last 8 feet, we must use formula (IV.). We have, then, $l - x = 5$; and

$$\text{Shear} = \frac{15000}{200}5 = 375 \text{ lbs.}$$

Whenever formulæ (I.) or (II.) apply, the second wheel is at the point: when (III.) or (IV.) apply, the first wheel is at the point.

11

TABLE I.

CONCENTRATED LOAD SYSTEM.

SHEAR.

l = length of span.

x = distance from left end, in feet, of any point at which the maximum shear is required.

Values of P and A to be taken from table for corresponding $l - x$, or $l - x - 8$.

(I.) From $x = 0$ to $x = 8$, second wheel at point, $\}$ Shear $= \dfrac{P}{l}(l - x) - \dfrac{A}{l}$

From $x = 8$ to limit, second wheel at point, $\}$ Shear $= \dfrac{p_1}{l}(l - x + 8) + \dfrac{P}{l}(l - x) - \dfrac{A}{l} - 15000.$ (II.)

Limit to be found from table on right, $p_1 = 15000.$

(III.) From $x = limit$ to $x = l - 8$, first wheel at point, $\}$ Shear $= \dfrac{p_1}{l}(l - x) + \dfrac{P}{l}(l - x - 8) - \dfrac{A}{l}.$

From $x = l - 8$ to $x = l$, first wheel at point, $\}$ Shear $= \dfrac{p_1}{l}(l - x).$ (IV.)

WEIGHTS, LBS.		Distance.	$l-x$ or $l-x-8.$	P	A	WEIGHTS, LBS.		Distances.	$l-x$ or $l-x-8.$	P	A	Span. feet	Limit. $x.$	Span. feet	Limit. $x.$
Locomotive.	$p_1 = 15000$	8					$p_{18} = 15000$	4	91			27	18.6	61	38.5
	$p_2 = 25000$	5	0	25000	0	Car.	$p_{19} = 12000$	5	95	335000	14370000	28	19	62	39¼
	$p_3 = 35000$	5	5	50000	125000		$p_{20} = 12000$		100	347000	15510000	29	19.4	63	39¼
	$p_4 = 25000$	5	10	75000	375000		$p_{21} = 12000$	10	110	359000	16710000	30	19.9	64	39¼
	$p_5 = 25000$	7	15	100000	750000		$p_{22} = 12000$	5	115	371000	18030000	31	20.6	65	39¼
Tender.	$p_6 = 15000$	5	22	115000	1080000	Car.	$p_{23} = 12000$	5	120	383000	19410000	32	21.3	66	39¼
	$p_7 = 15000$	5	27	130000	1485000		$p_{24} = 12000$	5	125	395000	20850000	33	22	67	39¼
	$p_8 = 15000$	9	32	145000	1965000		$p_{25} = 12000$	10	135	407000	22350000	34	22.7	68	40¼
	$p_9 = 15000$	8	37	160000	2520000	Car.	$p_{26} = 12000$	5	140	419000	23970000	35	23.4	69	40¼
Locomotive.	$p_{10} = 15000$	8	46	175000	3210000		$p_{27} = 12000$	5	145	431000	25650000	36	24.1	70	41¼
	$p_{11} = 25000$	5	54	200000	3960000	Car.	$p_{28} = 12000$	5	150	443000	27390000	37	24.8	71	41¼
	$p_{12} = 25000$	5	59	225000	6035000		$p_{29} = 12000$	10	160	455000	29190000	38	25.5	72	42¼
	$p_{13} = 25000$	5	64	250000	7635000		$p_{30} = 12000$	5	165	467000	31110000	39	26.2	73	42¼
	$p_{14} = 25000$	7	69	275000	9360000		$p_{31} = 12000$	5	170	479000	33090000	40	26.8	74	42¼
Tender.	$p_{15} = 15000$	5	76	290000	10500000	Car.	$p_{32} = 12000$	10	175	491000	35130000	41	27.2	75	42¼
	$p_{16} = 15000$	5	81	305000	11715000		$p_{33} = 12000$	5	185	503000	37230000	42	27.6	76	43¼
	$p_{17} = 15000$	5	86	320000	13005000		$p_{34} = 12000$	5	190	515000	39450000	43	28	77	43¼
	$p_{18} = 15000$		91			Car.	$p_{35} = 12000$		195	527000	41730000	44	28.7	78	44¼
												45	29.4	79	44¼
												46	30.1	80	45¼
												47	30.8	81	45¼
												48	31.5	82	45¼
												49	32.2	83	45¼
												50	32.9	84	46¼
												51	33.6	85	46¼
												52	34.3	86	47¼
												53	35	87	47¼
												54	35.4	88	48¼
												55	35.8	89	48¼
												56	36.2	90	48¼
												57	36.6	91	48¼
												58	37	92	48¼
												59	37.4	93	48¼
												60	37¾	94	47¼

TABLE I.

CONCENTRATED LOAD SYSTEM.

SHEAR.

l = length of span.

x = distance from left end, in feet, of any point at which the maximum shear is required.

Values of P and A to be taken from table for corresponding $l - x$, or $l - x - 8$.

(I.) $\quad$ From $x = o$ to $x = 8$, second wheel at point, $\quad$ Shear $= \dfrac{P}{l}(l - x) - \dfrac{A}{l}$.

From $x = 8$ to limit, second wheel at point, $\quad$ Shear $= \dfrac{p_1}{l}(l - x + 8) + \dfrac{P}{l}(l - x) - \dfrac{A}{l} - 15000$. $\quad$ (II.)

Limit to be found from table on right, $p_1 = 15000$.

(III.) $\quad$ From $x = limit$ to $x = l - 8$, first wheel at point, $\quad$ Shear $= \dfrac{p_1}{l}(l - x) + \dfrac{P}{l}(l - x - 8) - \dfrac{A}{l}$.

From $x = l - 8$ to $x = l$, first wheel at point, $\quad$ Shear $= \dfrac{p_1}{l}(l - x)$. $\quad$ (IV.)

Weights, lbs.	Distance	$l-x$ or $l-x-8$	P	A
$p_{25} = 12000$	0.	195		
$p_{26} = 12000$	5	200	539000	44070000
$p_{27} = 12000$	10	210	551000	46470000
$p_{28} = 12000$	5	215	563000	48990000
$p_{29} = 12000$	5	220	575000	51570000
$p_{30} = 12000$	5	225	587000	54210000
$p_{31} = 12000$	10	235	599000	56910000
$p_{32} = 12000$	5	240	611000	59730000
$p_{33} = 12000$	5	245	623000	62610000
$p_{34} = 12000$	-5	250	635000	65550000
$p_{35} = 12000$	10	260	647000	68550000
$p_{36} = 12000$	5	265	659000	71670000
$p_{37} = 12000$	5	270	671000	74850000
$p_{38} = 12000$	5	285	683000	78090000
$p_{39} = 12000$	10	285	695000	81390000
$p_{40} = 12000$	5	290	707000	84810000
$p_{41} = 12000$	5	295	719000	88290000
$p_{42} = 12000$		300	731000	91830000

Span. feet	Limit. x	Span. feet	Limit. x	Span. feet	Limit. x	Span. feet	Limit. x	Span. feet	Limit. x	Span. feet	Limit. x
95	47½	128	62.5	161	79¾	194	91¹¹⁄₁₆	227	89⁷⁄₁₆	260	91⅛
96	47½	129	63.2	162	79⅞	195	92⁷⁄₁₆	228	89¼	261	91¹⁄₁₆
97	47⅜	130	63.9	163	80⅛	196	92¼	229	89¾	262	91⅛
98	47⅜	131	64.6	164	80⅜	197	92⅜	230	89¾	263	91⅛
99	47⅜	132	65.3	165	81⅛	198	91⅛	231	89⅞	264	92⅛
100	47⅜	133	66	166	81¼	199	91⅞	232	90⅛	265	92⅛
101	47½	134	66.4	167	81⅜	200	89¼	233	89⅞	266	92⅛
102	47.6	135	66.8	168	81½	201	89⅜	234	90⅛	267	92⅞
103	48	136	67.2	169	82⅜	202	88½	235	90⅛	268	92⅛
104	48.4	137	67.8	170	82⅞	203	88⁷⁄₁₆	236	90⅛	269	93⁷⁄₁₆
105	48.8	138	68.5	171	83⅜	204	88⅜	237	91⁷⁄₁₆	270	93⅛
106	49.2	139	69.2	172	83⅞	205	89⁷⁄₁₆	238	91⅛	271	93⅛
107	49.6	140	69.9	173	84⅜	206	89⁵⁄₁₆	239	91⅛	272	94⅛
108	50	141	70.6	174	84½	207	90⅛	240	92⅛	273	93⅞
109	50.4	142	71.3	175	84⅝	208	90¼	241	92⅛	274	93⅞
110	50.9	143	72	176	84¾	209	90⅜	242	91⅛	275	93⅛
111	51.6	144	72.7	177	85⅝	210	89¼	243	92⁷⁄₁₆	276	93⅛
112	52.3	145	73.4	178	85⅞	211	90⅛	244	92⅛	277	91⅛
113	53	146	74.1	179	86⅜	212	90⅛	245	93⅛	280	90⅞
114	53.7	147	74.6	180	86⅝	213	91⅜	246	93⅜	281	90⅛
115	54.4	148	75	181	87⅛	214	91⅛	247	93⅛	285	92⁷⁄₁₆
116	55.1	149	75.4	182	87⅜	215	91¹⁄₁₆	248	93⅛	286	91⅛
117	55.8	150	75.8	183	87⅞	216	91⅛	249	93⅛	287	91⅛
118	56.5	151	76.2	184	88⅞	217	91⅛	250	92⅞	291	93⅛
119	57.2	152	76.6	185	88⅛	218	91⅛	251	90⁷⁄₁₆	292	93⅛
120	57.8	153	77	186	89⁷⁄₁₆	219	92⅜	252	90⅛	293	93⁷⁄₁₆
121	58.2	154	77.4	187	89½	220	92⅛	253	90⅛	294	93⁷⁄₁₆
122	58.6	155	78¼	188	90⁷⁄₁₆	221	93⅛	254	89⅞	295	93⁷⁄₁₆
123	59	156	78⅜	189	90½	222	92⅜	255	89⅛	296	94⅛
124	59.7	157	78⅝	190	90⅝	223	92⅜	256	90⅛	297	94⅛
125	60.4	158	78¾	191	90⅞	224	92⅛	257	90⅛	298	94⅛
126	61.1	159	78⅞	192	91⅛	225	92¹⁄₁₆	258	90⅛	299	94⁷⁄₁₂
127	61.8	160	78⅞	193	91¾	226	89¼	259	91⁷⁄₁₆	300	94⅛

iv

TABLE II.

CONCENTRATED LOAD SYSTEM.

* In applying II. to a *framed girder*, if the panel length is *more than 8 feet*, we must put instead of 15000 (= p_1), that *portion* of p_1 only, which, by the law of the lever, takes effect at the 1st apex on left of p_1.

TABLE I.

CONCENTRATED LOAD SYSTEM.

SHEAR.

l = length of span.

x = distance from left end, in feet, of any point at which the maximum shear is required.

Values of P and A to be taken from table for corresponding $l - x$, or $l - x - 8$.

(I.) *From $x = 0$ to $x = 8$, second wheel at point,* $\left\{ \text{Shear} = \frac{P}{l}(l - x) - \frac{A}{l}. \right.$

From $x = 8$ to limit, second wheel at point, $\left\{ \text{Shear} = \frac{p_1}{l}(l - x + 8) + \frac{P}{l}(l - x) - \frac{A}{l} - 15000. \right.$ (II.)

Limit to be found from table on right, $p_1 = 15000$.

(III.) *From $x =$ limit to $x = l - 8$, first wheel at point,* $\left\{ \text{Shear} = \frac{p_1}{l}(l - x) + \frac{P}{l}(l - x - 8) - \frac{A}{l}. \right.$

From $x = l - 8$ to $x = l$, first wheel at point, $\left\{ \text{Shear} = \frac{p_1}{l}(l - x). \right.$ (IV.)

Weights, lbs.	Distances	$l - x$ or $l - x - 8$.	P	A	Span.	Limit.	Span.	Limit.	Span.	Limit.	Span.	Limit.	Span.	Limit.	Span.	Limit.
					feet.	x.	feet.	x.	feet.	x.	feet.	x.	feet.	x.	feet.	x.

TABLE II.

CONCENTRATED LOAD SYSTEM.

MOMENTS.

TABLE II.

CONCENTRATED LOAD SYSTEM.
MOMENTS.

EXPLANATION OF TABLE.

FOR the maximum moment at any point of any span given in the table, simply multiply the value of the reaction R by the value of x, the distance of point from the left end, and subtract the value of m. The index of p, in the last column, shows the number of the wheel which must rest on the point, and hence the position of the load-system on the span. The load-system itself is given in Table 1.

Thus, for span of 200 feet : —

At 5 feet from left end, the reaction at left end is $R = 305175$ lbs. The moment is

$$Rx = 305175 \times 5 = 1525875.$$

The second wheel of the system must rest on the point.

At 10 feet from left, the reaction is $R = 306850$ lbs. The moment is

$$Rx - m = 306850 \times 10 - 120000 = 2948500.$$

The second wheel of the system must rest on the point.

At 20 feet from left, the reaction is $R = 293600$ lbs. The moment is

$$Rx - m = 293600 \times 20 - 320000 = 5552000.$$

The third wheel of the system must rest on the point.

To find the moment *at a point not given by the table*, find the reaction R for this point by interpolation, and then proceed as before. Thus,

At 30.7 feet from left, the reaction is $280650 - (280650 - 278060)\frac{7}{10} = 278837$. The moment is

$$Rx - m = 278837 \times 30.7 - 645000 = 7915296.$$

The fourth wheel of the system must rest on the point.

If, however, this point *falls between two "fields,"* find the *moment* by interpolation. Thus,

At 45.4 feet from left, we have the moment at 45, $255050 \times 45 - 1095000 = 10382250$. The moment at 46 is $270290 \times 46 - 1900000 = 10533340$. At 45.4, then, we have

$$10382250 + (10533340 - 10382250)\frac{4}{10} = 10442686.$$

The fifth or sixth wheel at point, indifferently.

In general, it will not be necessary to interpolate, but will be sufficient to take the nearest foot given in the table.

To find the moment at any point of a span not given in the table, find the moment *at the same point* for the two nearest spans given, and interpolate.

Thus, span 204 feet, moment at 10 feet from left : —

The moment at 10 feet from left, for a span of 200 feet, is 2948500. The moment at 10 feet from left, for span of 210 feet, is 3063330. The moment at 10 feet from left, for 204 feet span, is, then,

$$2948500 + (3063330 - 2948500)\frac{4}{10} = 2994432.$$

The second wheel of system at point.

TABLE II.

CONCENTRATED LOAD SYSTEM.

MOMENTS.

$$M_x = Rx - m.$$

SPAN, 10 feet.			SPAN, 11 feet.			SPAN, 12 feet.			SPAN, 13 feet.		
x	R	m	x	R	m	x	R	m	x	R	m
1	32500		1	34090		1	37500		1	40384		
2	27500		2	29545		2	31250		2	34615		
3	22500	0 f_2	3	25000	0 f_2	3	27083	0 f_2	3	28846	0 f_2
4	17500		4	20454		4	22916		4	25000		
5	12500		5	25909		5	18750		5	21153		
			6	21363							
						6	37500	125000 f_3	6	40384	125000 f_3
									7	34615	

SPAN, 14 feet.			SPAN, 15 feet.			SPAN, 16 feet.			SPAN, 17 feet.		
x	R	m	x	R	m	x	R	m	x	R	m
1	42857		1	45000		1	46875		1	50000		
2	37500		2	40000		2	42187		2	44117		
3	32142	0 f_2	3	35000	0 f_2	3	37500	0 f_2	3	39705	0 f_2
4	26785		4	30000		4	32812		4	35294		
5	23214		5	25000		5	28125		5	30882		
6	42857	125000 f_3	6	45000	125000 f_3	6	46875	125000 f_3	6	50000	125000 f_3
7	37500		7	40000		7	42187		7	44117	
			8	35000		8	37500		8	39705	
									9	35294	

MOMENTS.

$$M_x = Rx - m.$$

Span, 18 feet.			Span, 19 feet.			Span, 20 feet.			Span, 21 feet.		
x	R	m	x	R	m	x	R	m	x	R	m
1	52777		1	55263		1	57500		1	59523		
2	47222		2	50000		2	52500		2	54761		
3	41666	0 p_2	3	44736	0 p_2	3	47500	0 p_2	3	50000	0 p_2
4	37500		4	39473		4	42500		4	45238		
5	33333		5	35526		5	37500		5	40476		
6	52777		6	55263		6	57500		6	59523		
7	47222		7	50000		7	52500		7	54761		
8	41666	125000 p_3	8	44736	125000 p_3	8	47500	125000 p_3	8	50000	125000 p_3
9	37500		9	39473		9	42500		9	45238		
			10	35526		10	37500		10	40476		
									11	35714		

Span, 22 feet.			Span, 23 feet.			Span, 24 feet.			Span, 25 feet.		
x	R	m	x	R	m	x	R	m	x	R	m
1	61363		1	63043		1	65208		1	67200		
2	56818		2	58695		2	60416		2	62600		
3	52272	0 p_2	3	54847	0 p_2	3	56250	0 p_2	3	58000	0 p_2
4	47727		4	50000		4	52083		4	54000		
5	43181		5	45652		5	47915		5	50000		
6	61363		6	63043		6	65208		6	67200		
7	56818		7	58695		7	60416		7	62600		
8	52272	125000 p_3	8	54847	125000 p_3	8	56250	125000 p_3	8	58000	125000 p_3
9	47727		9	50000		9	52083		9	54000		
10	43181		10	45652		10	47915		10	50000		
11	38636		11	41304		11	43749		11	46000		
			12	36956		12	39582		12	42000		
									13	38000		

MOMENTS.

$$M_x = Rx - m.$$

SPAN, 26 feet.

x	R	m	
1	69038		
2	64615		
3	60192	0	p₂
4	55769		
5	51923		
6	69038		
7	64615		
8	60192		
9	55769	125000	p₃
10	51923		
11	48076		
12	44230		
13	40384		

SPAN, 27 feet.

x	R	m	
1	70740		
2	66481		
3	62222	0	p₂
4	57962		
5	53703		
6	50000		
7	66481		
8	62222		
9	57962	125000	p₃
10	53703		
11	50000		
12	46296		
13	42592		
14	53333	330000	p₃

SPAN, 28 feet.

x	R	m	
1	72321		
2	68213		
3	64106	0	p₂
4	60000		
5	55891		
6	51783		
7	68213		
8	64106		
9	60000	125000	p₃
10	55891		
11	51783		
12	48212		
13	44640		
14	55533	330000	p₃

SPAN, 29 feet.

x	R	m	
1	74306		
2	69820		
3	65855	0	p₂
4	61889		
5	57924		
6	53958		
7	69820		
8	65855		
9	61889	125000	p₃
10	57924		
11	53958		
12	49993		
13	46544		
14	57578	330000	p₃
15	53613		

SPAN, 30 feet.

x	R	m	
1	76166		
2	71833		
3	67500	0	p₂
4	63666		
5	59833		
6	56000		
7	71833		
8	67500		
9	63666	125000	p₃
10	59833		
11	56000		
12	52166		
13	48333		
14	59500	330000	p₃
15	55666		

SPAN, 31 feet.

x	R	m	
1	77903		
2	73709		
3	69516	0	p₂
4	65322		
5	61612		
6	57903		
7	73709		
8	69516		
9	65322	125000	p₃
10	61612		
11	57903		
12	54193		
13	50483		
14	61290	330000	p₃
15	57580		
16	53870		

SPAN, 32 feet.

x	R	m	
1	79531		
2	75468		
3	71406	0	p₂
4	67343		
5	63281		
6	59687		
7	75468		
8	71406		
9	67343	125000	p₃
10	63281		
11	59687		
12	56093		
13	52500		
14	63436	330000	p₃
15	59373		
16	55780		

SPAN, 33 feet.

x	R	m	
1	81060		
2	77121		
3	73181	0	p₂*
4	69242		
5	65303		
6	61363		
7	77121		
8	73181		
9	69242	125000	p₃
10	65303		
11	61363		
12	57878		
13	54393		
14	65454	330000	p₃
15	61514		
16	57574		
17	54090		

MOMENTS.

$$M_x = Rx - m.$$

SPAN, 34 feet.

x	R	m	
1	82941		
2	78676		
3	74852	0	p_2
4	71029		
5	67205		
6	63382		
7	78676		
8	74852		
9	71029		
10	67205	125000	p_3
11	63382		
12	59558		
13	56176		
14	67353		
15	63528	320000	p_3
16	59705		
17	55822		

SPAN, 35 feet.

x	R	m	
1	84714		
2	80571		
3	76428	0	p_2
4	72714		
5	69000		
6	65285		
7	80571		
8	76428		
9	72714		
10	69000	125000	p_3
11	65285		
12	61571		
13	57857		
14	69143		
15	65428		
16	61714	320000	p_3
17	58000		
18	54286		

SPAN, 36 feet.

x	R	m	
1	86388		
2	82361		
3	78333	0	p_2
4	74305		
5	70694		
6	67083		
7	82361		
8	78333		
9	74305		
10	70694	125000	p_3
11	67083		
12	63472		
13	59861		
14	70833		
15	67221		
16	63611	320000	p_3
17	60000		
18	56388		

SPAN, 37 feet.

x	R	m	
1	87972		
2	84054		
3	80135	0	p_2
4	76216		
5	72297		
6	68702		
7	84054		
8	80135		
9	76216		
10	72297	125000	p_3
11	68702		
12	65108		
13	61513		
14	72513		
15	68513		
16	65000	320000	p_3
17	61486		
18	57937		
19	54460		

SPAN, 38 feet.

x	R	m	
1	89473		
2	85657		
3	81842	0	p_4
4	78027		
5	74211		
6	70395		
7	85657		
8	81842		
9	78027		
10	74211	125000	p_5
11	70395		
12	66974		
13	63553		

SPAN, 38 feet.

x	R	m	
14	74737		
15	70921		
16	67105	320000	p_5
17	63685		
18	60263		
19	74737	645000	p_4

SPAN, 39 feet.

x	R	m	
1	91282		
2	87179		
3	83461	0	p_4
4	79743		
5	76025		
6	72307		
7	87179		
8	83461		
9	79743		
10	76025	125000	p_5
11	72307		
12	68589		
13	65256		

SPAN, 39 feet.

x	R	m	
14	76538		
15	72819		
16	69102	320000	p_5
17	65384		
18	62050		
19	76538	645000	p_4

MOMENTS.

$$M_x = Rx - m.$$

Span, 40 feet			Span, 41 feet			Span, 42 feet			Span, 43 feet		
x	R	m	x	R	m	x	R	m	x	R	m
1	93000		1	94634		1	96190		1	97674		
2	89000		2	90731		2	92380		2	93953		
3	89000	0	3	86829	0	3	88571	0	3	90232	0	
4	81375	p_2	4	82926	p_2	4	84761	p_2	4	86511	p_2
5	77750		5	79390		5	80952		5	82790		
6	74125		6	75853		6	77500		6	79069		
7	89000		7	90731		7	92380		7	93953		
8	85000		8	86829		8	88571		8	90232		
9	81375		9	82926		9	84761		9	86511		
10	77750	125000	10	79390	125000	10	80952	125000	10	82790	125000
11	74125	p_3	11	75853	p_3	11	77500	p_3	11	79069	p_3
12	70500		12	72317		12	74047		12	75697		
13	66875		13	68780		13	70595		13	72325		
14	78250		14	79877		14	81784		14	83604		
15	74625	320000	15	76341	320000	15	77975	320000	15	79883	320000
16	71000	p_3	16	72804	p_3	16	74523	p_3	16	76162	p_3
17	67375		17	69267		17	71071		17	72790		
			18	65730		18	67618		18	69417		
18	81875		19	79877		19	81784		19	83604		
19	78250	645000	20	76341	645000	20	77975	645000	20	79883	645000
20	74625	p_4	21	72804	p_4	21	74523	p_4	21	76162	p_4
									22	72790		

Span, 44 feet			Span, 44 feet			Span, 45 feet			Span, 45 feet		
x	R	m	x	R	m	x	R	m	x	R	m
1	99090		14	85340		1	100444		14	87000		
2	95454		15	81704	320000	2	96888		15	83444	320000
3	91818	0	16	78067	p_3	3	93333	0	16	80000	p_3
4	88181	p_2	17	74431		4	89777	p_2	17	76333	
5	84545		18	71135		5	86222		18	72777		
6	80909					6	82666					
			19	85340		7	79111					
7	95454		20	81704	645000				19	87000		
8	91818		21	78067	p_4	8	93333		20	83444	645000
9	88181		22	74431		9	89777		21	80000	p_4
10	84545	125000				10	86222	125000	22	76333	
11	80909	p_3				11	82666	p_3	23	72777	
12	77272					12	79111					
13	73977					13	75555					

MOMENTS.

$$M_x = Rx - m.$$

x	SPAN, 46 feet. R	m		x	SPAN, 47 feet. R	m		x	SPAN, 48 feet. R	m		x	SPAN, 49 feet. R	m	
1	101739			1	102978			1	104479			1	105918		
2	98260			2	99574			2	100834			2	102346		
3	94782			3	96170			3	97500			3	98775		
4	91304	0	p_2	4	92765	0	p_2	4	94166	0	p_2	4	95509	0	p_2
5	87826			5	89361			5	90833			5	92244		
6	84347			6	85957			6	87500			6	88978		
7	80869			7	82553			7	84166			7	85712		
8	94782			8	96170			8	97500			8	98775		
9	91304			9	92765			9	94166			9	95509		
10	87826			10	89361			10	90833			10	92244		
11	84347	125000	p_3	11	85957	125000	p_3	11	87500	125000	p_3	11	88978	125000	p_3
12	80869			12	82553			12	84166			12	85712		
13	77391			13	79148			13	80833			13	82446		
14	88586			14	90424			14	92187			14	93874		
15	85107			15	86701			15	88541			15	90302		
16	81629	320000	p_3	16	83297	320000	p_3	16	84895	320000	p_3	16	86729	320000	p_3
17	78151			17	79893			17	81563			17	83159		
18	74673			18	76489			18	78229			18	79893		
19	88586			19	90424			19	92187			19	93874		
20	85107			20	86701			20	88541			20	90302		
21	81629			21	83297			21	84895			21	86729		
22	78151	645000	p_4	22	79893	645000	p_4	22	81563	645000	p_4	22	83159	645000	p_4
23	74673			23	76489			23	78229			23	79893		
				24	73085			24	74896			24	76628		
												25	73363		

x	SPAN, 50 feet. R	m		x	SPAN, 50 feet. R	m		x	SPAN, 51 feet. R	m		x	SPAN, 51 feet. R	m	
1	107300			14	95500			1	108627			14	97057		
2	103300			15	92000			2	105196			15	93626		
3	100300			16	88500	320000	p_3	3	101764			16	90195	320000	p_3
4	96800	0	p_2	17	85000			4	98333	0	p_2	17	86764		
5	93600			18	81500			5	94901			18	83332		
6	90400							6	91764						
7	87200			19	95500			7	88627			19	97057		
				20	92000							20	93626		
8	100300			21	88500			8	101764			21	90195		
9	96800			22	85000			9	98333			22	86764		
10	93600	125000	p_3	23	81500	645000	p_4	10	94901	125000	p_3	23	83332	645000	p_4
11	90400			24	78300			11	91764			24	79901		
12	87200			25	75100			12	88627			25	76764		
13	84000							13	85490			26	73627		

MOMENTS.

$$M_x = Rx - m.$$

Span, 52 feet.			Span, 53 feet.			Span, 54 feet.			Span, 55 feet.		
x	R	m	x	R	m	x	R	m	x	R	m
1	109903		1	111132		1	112314		1	113451		
2	106538		2	107830		2	109074		2	110272		
3	103173	0	3	104528	0	3	105833	0	3	107090	0
4	99807	p_2	4	101226	p_2	4	102592	p_2	4	103909	p_2
5	96442		5	97924		5	99351		5	100727		
6	93076		6	94622		6	96111		6	97545		
7	90000		7	91320		7	92870		7	94363		
8	103173		8	104528		8	105833		8	107090		
9	99807		9	101226		9	102592		9	103909		
10	96442	125000	10	97924	125000	10	99351	125000	10	100727	125000
11	93076	p_3	11	94622	p_3	11	96111	p_3	11	97545	p_3
12	90000		12	91320		12	92870		12	94363		
13	86923		13	88301		13	89629		13	91181		
14	98557		14	100000		14	101388		14	102727		
15	95192		15	96697		15	98147		15	99544		
16	91826	320000	16	93395	320000	16	94906	320000	16	96362	320000
17	88461	p_3	17	90093	p_3	17	91665	p_3	17	93181	p_3
18	85095		18	86791		18	88425		18	90000		
19	81730		19	83489		19	85184		19	86817		
						20	81943		20	83635		
20	95192		20	96697		21	94906		21	96362		
21	91826		21	93395		22	91665		22	93181		
22	88461		22	90093		23	88425		23	90000		
23	85095	640000	23	86791	640000	24	85184	640000	24	86817	640000
24	81730	p_4	24	83489	p_4	25	81943	p_4	25	83635	p_4
25	78364		25	80187		26	78702		26	80454		
26	75288		26	76885		27	75462		27	77272		
			27	73867					28	74090		

Span, 56 feet.			Span, 56 feet.			Span, 56 feet.			Span, 56 feet.		
x	R	m	x	R	m	x	R	m	x	R	m
1	115000		8	108303		14	104285		21	97767		
2	111428		9	105178		15	100892		22	94642		
3	108303	0	10	102053	125000	16	97767	320000	23	91517	640000
4	105178	p_2	11	98928	p_3	17	94642	p_3	24	88392	p_4
5	102053		12	95803		18	91517		25	85267		
6	98928		13	92678		19	88392		26	82142		
7	95803					20	85267		27	79017		
									28	75892		

MOMENTS.

$$M_x = Rx - m.$$

SPAN, 57 feet.				SPAN, 58 feet.				SPAN, 59 feet.				SPAN, 60 feet.			
x	R	m		x	R	m		x	R	m		x	R	m	
1	116491			1	117931			1	119322			1	120666		
2	112983			2	114483			2	115932			2	117333		
3	109475			3	111034			3	112542			3	114000		
4	106405	0	p_2	4	107586	0	p_2	4	109152	0	p_2	4	110666	0	p_2
5	103335			5	104569			5	105762			5	107333		
6	100265			6	101552			6	102796			6	104000		
7	97195			7	98534			7	99830			7	101083		
8	109475			8	111034			8	112542			8	114000		
9	106405			9	107586			9	109152			9	110666		
10	103335	125000	p_3	10	104569	125000	p_3	10	105762	125000	p_3	10	107333	125000	p_3
11	100265			11	101552			11	102796			11	104000		
12	97195			12	98534			12	99830			12	101083		
13	94124			13	95517			13	96864			13	98166		
14	105790			14	107241			14	108643			14	110000		
15	102457			15	103965			15	105423			15	106833		
16	99124			16	100689			16	102203			16	103666		
17	96054	320000	p_3	17	97413	320000	p_3	17	98983	320000	p_3	17	100500	320000	p_3
18	92984			18	94396			18	95761			18	97333		
19	89914			19	91379			19	92796			19	94166		
20	86843			20	88361			20	89830			20	91250		
21	83773			21	85345			21	86864			21	88333		
22	96054			22	97413			22	98983			22	100500		
23	92984			23	94396			23	95761			23	97333		
24	89914			24	91379			24	92796			24	94166		
25	86843	645000	p_4	25	88361	645000	p_4	25	89830	645000	p_4	25	91250	645000	p_4
26	83773			26	85345			26	86864			26	88333		
27	80703			27	82327			27	83897			27	85416		
28	77633			28	79309			28	80931			28	82500		
29	74563			29	76293			29	77965			29	79583		
								30	74500			30	76666		

MOMENTS.

$$M_x = Rx - m.$$

	SPAN, 61 feet.			SPAN, 62 feet.			SPAN, 63 feet.			SPAN, 64 feet.	
x	R	m		x	R	m		x	R	m		x	R	m	
1	122377			1	124032			1	125634			1	127187		
2	118688			2	120403			2	122003			2	123671		
3	115409			3	116774			3	118492			3	120156		
4	112131	0	p_2	4	113548	0	p_2	4	114920	0	p_2	4	116640	0	p_2
5	108852			5	110322			5	111746			5	113125		
6	105573			6	107096			6	108571			6	110000		
7	102295			7	103870			7	105396			7	106875		
8	115409			8	116774			8	118492			8	120156		
9	112131			9	113548			9	114920			9	116640		
10	108852	125000	p_3	10	110322	125000	p_3	10	111746	125000	p_3	10	113125	125000	p_3
11	105573			11	107096			11	108571			11	110000		
12	102295			12	103870			12	105396			12	106875		
13	99426			13	100644			13	102222			13	103750		
14	111311			14	112580			14	113808			14	115390		
15	108196			15	109516			15	110792			15	112031		
16	105081			16	106451			16	107777			16	109061		
17	101966	320000	p_3	17	103386	320000	p_3	17	104761	320000	p_3	17	106093	320000	p_3
18	98851			18	100322			18	101745			18	103124		
19	95737			19	97257			19	98729			19	100156		
20	92622			20	94193			20	95713			20	97188		
21	89753			21	91128			21	92698			21	94219		
22	101966			22	103386			22	104761			22	106093		
23	98851			23	100322			23	101745			23	103124		
24	95737			24	97257			24	98729			24	100156		
25	92622			25	94193			25	95713			25	97188		
26	89753	645000	p_4	26	91128	645000	p_4	26	92698	645000	p_4	26	94219	645000	p_4
27	86635			27	88365			27	89682			27	91250		
28	84016			28	85483			28	86904			28	88281		
29	81148			29	82660			29	84126			29	85546		
30	78279			30	79837			30	81348			30	82812		
31	75410			31	77015			31	78570			31	80078		
								32	75792			32	77343		

MOMENTS.

$$M_x = Rx - m.$$

Span, 65 feet.				Span, 66 feet.				Span, 67 feet.				Span, 68 feet.		
x	R	m		x	R	m		x	R	m		x	R	m
1	128692			1	130530			1	132313			1	134044	
2	125231			2	126742			2	128582			2	130367	
3	121769	0	p_2	3	123333	0	p_2	3	124850	0	p_2	3	126691	0	p_2
4	118308			4	119924			4	121492			4	123014	
5	114846			5	116515			5	118134			5	119706	
6	111385			6	113106			6	114776			6	116398	
7	108308			7	109696			7	111417			7	113089	
8	121769			8	123333			8	124850			8	126691	
9	118308			9	119924			9	121492			9	123014	
10	114846	125000	p_3	10	116515	125000	p_3	10	118134	125000	p_3	10	119706	125000	p_3
11	111385			11	113106			11	114776			11	116398	
12	108308			12	109696			12	111417			12	113089	
13	105230			13	106666			13	108059			13	109780	
14	116923			14	118408			14	119850			14	121250	
15	113615			15	115151			15	116641			15	118088	
16	110307			16	111893			16	113432			16	114927	
17	107384			17	108635			17	110223			17	111765	
18	104461	320000	p_3	18	105756	320000	p_3	18	107014	320000	p_3	18	108603	320000	p_3
19	101538			19	102878			19	104178			19	105441	
20	98614			20	100000			20	101342			20	102647	
21	95691			21	97119			21	98506			21	99853	
22	92769			22	94241			22	95671			22	97055	
23	104461			23	105756			23	107014			23	108603	
24	101538			24	102878			24	104178			24	105441	
25	98614			25	100000			25	101342			25	102647	
26	95691			26	97119			26	98506			26	99853	
27	92769	645000	p_4	27	94241	645000	p_4	27	95671	645000	p_4	27	97055	645000	p_4
28	89847			28	91363			28	92835			28	94220	
29	86924			29	88484			29	90000			29	91471	
30	84232			30	85705			30	87163			30	88656	
31	81538			31	82954			31	84327			31	85863	
32	78845			32	80302			32	81715			32	83068	
33	89847	1095000	p_5	33	91363	1095000	p_5	33	92835	1095000	p_5	33	94220	1095000	p_5
								34	90000			34	91471	

MOMENTS.

$$M_x = Rx - m.$$

Span, 69 feet			Span, 70 feet			Span, 71 feet			Span, 72 feet		
x	R	m	x	R	m	x	R	m	x	R	m
1	135724		1	137357		1	139295		1	141180	
2	132101		2	133785		2	135422		2	137361	
3	128478	0	3	130214	0	3	131901	0	3	133541	0
4	124855	f_2	4	126642	f_2	4	128380	f_2	4	130070	f_2
5	121231		5	123071		5	124859		5	126600	
6	117971		6	119500		6	121338		6	123129	
7	114710		7	116285		7	117816		7	119658	
8	128478		8	130214		8	131901		8	133541	
9	124855		9	126642		9	128380		9	130070	
10	121231	125000	10	123071	125000	10	124859	125000	10	126600	125000
11	117971	P_3	11	119500	P_3	11	121338	P_3	11	123129	P_3
12	114710		12	116285		12	117816		12	119658	
13	111449		13	113071		13	114646		13	116187	
14	122970		14	124641		14	126264		14	127853	
15	119492		15	121212		15	122882		15	124520	
16	116375		16	117784		16	119501		16	121187	
17	113260	320000	17	114712	320000	17	116118	320000	17	117853	320000
18	110144	P_3	18	111641	P_3	18	113090	P_3	18	114520	P_3
19	107028		19	108570		19	110062		19	111534	
20	103912		20	105500		20	107035		20	108547	
21	101158		21	102437							
22	113260		22	114712		21	119501		21	121187	
23	110144		23	111641		22	116118		22	117853	
24	107028		24	108570		23	113090		23	114520	
25	103912		25	105500		24	110062		24	111534	
26	101158		26	102427		25	107035		25	108547	
27	98405	640000	27	99713	640000	26	104006	640000	26	105562	640000
28	95652	P_4	28	97000	P_4	27	100978	P_4	27	102576	P_4
29	92897		29	94284		28	98302		28	99589	
30	90144		30	91570		29	95626		29	96951	
31	87390		31	88857		30	92949		30	94312	
32	84637		32	86143		31	90274		31	91672	
			33	83428		32	87598		32	89039	
						33	84921		33	86395	
33	95652	1095000	34	94284	1095000	34	95626	1095000	34	96951	1095000
34	92897	P_5	35	91570	P_5	35	92949	P_5	35	94312	P_5
35	90144					36	90274		36	91672	

MOMENTS.

$$M_x = Rx - m.$$

Span, 73 feet.

x	R	m	p
1	143013		
2	139246		
3	135479		
4	131712	0	p_2
5	128287		
6	124863		
7	121438		
8	135479		
9	131712		
10	128287	125000	p_3
11	124863		
12	121438		
13	118013		
14	129383		
15	126095		
16	122807		
17	119520	320000	p_3
18	116232		
19	112945		
20	110000		
21	122807		
22	119520		
23	116232		
24	112945		
25	110000		
26	107054		
27	104108	649000	p_4
28	101164		
29	98218		
30	95616		
31	93012		
32	90410		
33	87807		
34	98218		
35	95616	1095000	p_5
36	93012		
37	90410		

Span, 74 feet.

x	R	m	p
1	144797		
2	141081		
3	137365	0	p_2
4	133649		
5	129933		
6	126555		
7	141081		
8	137365		
9	133649		
10	129933	125000	p_3
11	126555		
12	123176		
13	119798		
14	131216		
15	127635		
16	124391		
17	121148	320000	p_3
18	117904		
19	114661		
20	111418		
21	124391		
22	121148		
23	117904		
24	114661		
25	111418		
26	108513		
27	105607	649000	p_4
28	102701		
29	99797		
30	96891		
31	94323		
32	91756		
33	89188		
34	99797		
35	96891	1095000	p_5
36	94323		
37	91756		

Span, 75 feet.

x	R	m	p
1	146533		
2	142866		
3	139200	0	p_2
4	135533		
5	131866		
6	128200		
7	142866		
8	139200		
9	135533		
10	131866	125000	p_3
11	128200		
12	124866		
13	121533		
14	133000		
15	129466		
16	125933	320000	p_3
17	122733		
18	119533		
19	116333		
20	129466		
21	125933		
22	122733		
23	119533		
24	116333		
25	113133		
26	109933	649000	p_4
27	106666		
28	104200		
29	101333		
30	98466		
31	95600		
32	93066		
33	104200		
34	101333		
35	98466	1095000	p_5
36	95600		
37	93066		
38	90533		

Span, 76 feet.

x	R	m	p
1	148223		
2	144605		
3	140986	0	p_2
4	137368		
5	133750		
6	130131		
7	144605		
8	140986		
9	137368		
10	133750	125000	p_3
11	130131		
12	126513		
13	123223		
14	134776		
15	131249		
16	127762	320000	p_3
17	124266		
18	121118		
19	117959		
20	131249		
21	127762		
22	124266		
23	121118		
24	117959		
25	114802		
26	111644	649000	p_4
27	108486		
28	105657		
29	102828		
30	100000		
31	97170		
32	108486		
33	105657		
34	102828		
35	100000	1095000	p_5
36	97170		
37	94341		
38	91841		

MOMENTS.

$$M_x = Rx - m.$$

	Span, 77 feet.				Span, 78 feet.				Span, 79 feet.				Span, 80 feet.		
x	R	m		x	R	m		x	R	m		x	R	m	
1	149870			1	151666			1	153417			1	155125		
2	146298			2	147948			2	149746			2	151500		
3	143727	0	p_2	3	144423	0	p_2	3	146075	0	p_2	3	147875	0	p_2
4	139155			4	140897			4	142594			4	144250		
5	135584			5	137371			5	139113			5	140812		
6	132013			6	133845			6	135632			6	137375		
7	128442			7	130319			7	132151			7	133937		
8	142727			8	144423			8	146075			8	147875		
9	139155			9	140897			9	142594			9	144250		
10	135584	125000	p_3	10	137371	125000	p_3	10	139113	125000	p_3	10	140812	125000	p_3
11	132013			11	133845			11	135632			11	137375		
12	128442			12	130319			12	132151			12	133937		
13	124871			13	126793			13	128670			13	130500		
14	136429			14	138074			14	140000			14	141874		
15	132987			15	134677			15	136328			15	138250		
16	129545	320000	p_3	16	131280	320000	p_3	16	132974	320000	p_3	16	134624	320000	p_3
17	126104			17	127882			17	129620			17	131312		
18	122662			18	124484			18	126265			18	128000		
19	119546			19	121088										
20	132987			20	134677			19	140000			19	141874		
21	129545			21	131280			20	136328			20	138250		
22	126104			22	127882			21	132974			21	134624		
23	122662			23	124484			22	129620			22	131312		
24	119546	645000	p_4	24	121088	645000	p_4	23	126265	645000	p_4	23	128000	645000	p_4
25	116429			25	118010			24	122910			24	124687		
26	113312			26	114933			25	119556			25	121374		
27	110195			27	111857			26	116518			26	118062		
28	107077			28	108780			27	113480			27	115062		
29	104286			29	105702			28	110442			28	112062		
30	101494							29	107404						
				30	118010			30	119556			29	124687		
31	113312			31	114933			31	116518			30	121374		
32	110195			32	111857			32	113480			31	118062		
33	107077			33	108780			33	110442			32	115062		
34	104286	1095000	p_5	34	105702	1095000	p_5	33	110442	1095000	p_5	33	112062	1095000	p_5
35	101494			35	102946			34	107404			34	109062		
36	98701			36	100190			35	104366			35	106062		
37	95909			37	97433			36	101644			36	103062		
38	93116			38	94677			37	98923			37	100375		
39	90650			39	91923			38	96201			38	97687		
								39	93480			39	95000		
								40	90759			40	92312		

MOMENTS.

$$M_x = Rx - m.$$

Span, 81 feet.

x	R	m
1	156790	
2	153209	
3	149629	
4	146049	0 f_2
5	142469	
6	139074	
7	135679	
8	149629	
9	146049	
10	142469	
11	139074	125000 f_3
12	135679	
13	132283	
14	143702	
15	140122	
16	136542	310000 f_3
17	132962	
18	129691	
19	143702	
20	140122	
21	136542	
22	132962	
23	129691	649000 f_4
24	126418	
25	123147	
26	119876	
27	116604	
28	113641	
29	126418	
30	123147	
31	119876	
32	116604	
33	113641	
34	110678	1095000 f_5
35	107715	
36	104752	
37	101789	
38	99135	
39	96481	
40	93826	
41	91172	

Span, 82 feet.

x	R	m
1	158414	
2	154878	
3	151341	
4	147804	0 f_2
5	144268	
6	140731	
7	137378	
8	151341	
9	147804	
10	144268	
11	140731	125000 f_3
12	137378	
13	134024	
14	145487	
15	141951	
16	138414	310000 f_3
17	134877	
18	131314	
19	145487	
20	141951	
21	138414	
22	134877	
23	131341	649000 f_4
24	128109	
25	124877	
26	121645	
27	118413	
28	131341	
29	128109	
30	124877	
31	121645	
32	118413	
33	115182	1095000 f_5
34	112255	
35	109328	
36	106401	
37	103475	
38	100548	
39	97926	
40	95304	
41	92682	

Span, 83 feet.

x	R	m
1	160180	
2	156506	
3	153012	
4	149518	0 f_2
5	146024	
6	142530	
7	139036	
8	153012	
9	149518	
10	146024	
11	142530	125000 f_3
12	139036	
13	135722	
14	147228	
15	143734	
16	140239	310000 f_3
17	136746	
18	133251	
19	147228	
20	143734	
21	140239	
22	136746	649000 f_4
23	133251	
24	129757	
25	126564	
26	123372	
27	136746	
28	133251	
29	129757	
30	126564	1095000 f_5
32	120179	
34	113794	
36	108010	
38	102227	
39	120179	1900000 f_6
40	133251	
41	129757	2550000 f_7
42	126564	

Span, 84 feet.

x	R	m
1	161904	
2	158273	
3	154642	
4	151190	0 f_2
5	147738	
6	144285	
7	140833	
8	154642	
9	151190	
10	147738	
11	144285	125000 f_3
12	140833	
13	137380	
14	148938	
15	145475	
16	142023	310000 f_3
17	138570	
18	135118	
19	148938	
20	145475	
21	142023	
22	138570	649000 f_4
23	135118	
24	131666	
25	128214	
26	125059	
27	138570	
28	135118	
30	128214	
32	121903	1095000 f_5
34	115594	
35	112440	
36	109582	
37	106726	
38	125059	1900000 f_6
39	121903	
40	135118	
41	131666	2550000 f_7
42	128214	

MOMENTS.

$$M_x = Rx - m.$$

Span, 85 feet			Span, 86 feet			Span, 87 feet			Span, 88 feet		
x	R	m	x	R	m	x	R	m	x	R	m
1	163588		1	165232		1	166839		1	168579	
2	160000		2	161685		2	163333		2	164943	
3	156412		3	158139		3	159827		3	161477	
4	152824	0 p_2	4	154593	0 p_2	4	156321	0 p_2	4	158011	0 p_2
5	149412		5	151046		5	152816		5	154545	
6	146000		6	147674		6	149310		6	151079	
7	142588		7	144302		7	145977		7	147613	
8	156412		8	158139		8	159827		8	161477	
9	152824		9	154593		9	156321		9	158011	
10	149412	125000 p_3	10	151046	125000 p_3	10	152816	125000 p_2	10	154545	125000 p_3
11	146000		11	147674		11	149310		11	151079	
12	142588		12	144302		12	145977		12	147613	
13	139177		13	140930		13	142643		13	144317	
14	150387		14	152383		14	154137		14	155951	
15	147175		15	148837		15	150632		15	152386	
16	143763	320000 p_3	16	145464	320000 p_3	16	147125	320000 p_3	16	148919	320000 p_3
17	140351		17	142092		17	143792		17	145454	
18	136939		18	138720		18	140458		18	142158	
19	133528		19	135348		19	137125		19	138863	
20	147175		20	148837		20	150632		20	152386	
21	143763		21	145404		21	147125		21	148919	
22	140351		22	142092		22	143792		22	145454	
23	136939	645000 p_4	23	138720	645000 p_4	23	140458	645000 p_4	23	142158	645000 p_4
24	133528		24	135348		24	137125		24	138863	
25	130115		25	131976		25	133793		25	135567	
26	126704		26	128604		26	130459		26	132272	
						27	127126		27	128976	
27	140351		27	142092		28	140458		28	142158	
28	136939		28	138720		29	137125		29	138863	
29	133528		29	135348		30	133793		30	135567	
30	130115		30	131976		31	130459		31	132272	
31	126704	1095000 p_5	31	128604	1095000 p_5	32	127126	1095000 p_5	32	128976	1095000 p_5
32	123586		32	125232		33	123792		33	125680	
33	120469		33	122150		34	120746		34	122385	
34	117351		34	119068		35	117701		35	119373	
35	114233		35	115986							
36	133528		36	135348		36	137125		36	138863	
37	130115	1900000 p_6	37	131976	1900000 p_6	37	133793	1900000 p_6	37	135567	1900000 p_6
38	126704		38	128604		38	130459		38	132272	
						39	127126		39	128976	
39	140351		39	142092		40	140458		40	142158	
40	136939		40	138720		41	137125		41	138863	
41	133528	2550000 p_7	41	135348	2550000 p_7	42	133793	2550000 p_7	42	135567	2550000 p_7
42	130115		42	131976		43	130459		43	132272	
43	126704		43	128604		44	127126		44	128976	

MOMENTS.

$$M_x = Rx - m.$$

Span, 89 feet			Span, 89 feet			Span, 90 feet			Span, 90 feet		
x	R	m	x	R	m	x	R	m	x	R	m
1	170280		40	143819		1	171943		40	145610	
2	166685		41	140561	255000	2	168388		41	142222	255000
3	163089	0	42	137303	p_7	3	164832	0	42	139000	p_7
4	159662	p_2	43	134044		4	161277	p_2	43	135777	
5	156235		44	130786		5	157888		44	132555	
6	152808					6	154500				
7	149382					7	151110				
8	163089		45	143819	327500	8	164832		45	145610	327500
9	159662				p_8	9	161277				p_8
10	156235	125000				10	157888	125000			
11	152808	p_3				11	154500	p_3			
12	149382					12	151110				
13	145955					13	147721				
14	157527					14	159166				
15	154100					15	155776				
16	150673	320000				16	152388	320000			
17	147246	p_3				17	149000	p_3			
18	143819					18	145610				
19	140561					19	142222				
20	154100					20	155776				
21	150673					21	152388				
22	147246					22	149000				
23	143819	640000				23	145610	640000			
24	140561	p_4				24	142222	p_4			
25	137303					25	139000				
26	134044					26	135777				
27	130786					27	132555				
28	143819					28	145610				
29	140561					29	142222				
30	137303					30	139000				
31	134044	1095000				31	135777	1095000			
32	130786	p_5				32	132555	p_5			
33	127527					33	129333				
34	124269					34	126110				
35	121010					35	122888				
36	140561					36	142222				
37	137303	1900000				37	139000	1900000			
38	134044	p_6				38	135777	p_6			
39	130786					39	132555				

MOMENTS.

$$M_x = Rx - m$$

	SPAN, 91 feet				SPAN, 91 feet				SPAN, 92 feet				SPAN, 92 feet		
x	R	m		x	R	m		x	R	m		x	R	m	
1	173571			28	147361			1	175163			28	149075		
2	170054			29	144011			2	171684			29	145760		
3	166538	0	p_2	30	140660	1095000	p_5	3	168206	0	p_2	30	142444	1095000	p_5
4	163021			31	137473			4	164728			31	139129		
5	159505			32	134286			5	161250			32	135977		
6	156153			33	131100			6	157771			33	132825		
7	152802			34	127912			7	154456			34	129673		
				35	124725							35	126521		
8	166538			36	144011			8	168206			36	145760		
9	163021			37	140660			9	164728			37	142444		
10	159505	125000	p_3	38	137473	1900000	p_6	10	161250	125000	p_3	38	139129	1900000	p_6
11	156153			39	134286			11	157771			39	135977		
12	152802							12	154456						
13	149450							13	151141						
				40	147361							40	149075		
14	160933			41	144011			14	162662			41	145760		
15	157417			42	140660	2550000	p_7	15	159183			42	142444	2550000	p_7
16	154065	320000	p_3	43	137473			16	155705	320000	p_3	43	139129		
17	150713			44	134286			17	152390			44	135977		
18	147361							18	149075						
19	144011							19	145760						
				45	147361	3275000	p_8					45	149075	3275000	p_8
				46	144011							46	145760		
20	157417							20	159183						
21	154065							21	155705						
22	150713							22	152390						
23	147361	645000	p_4					23	149075	645000	p_4				
24	144011							24	145760						
25	140660							25	142444						
26	137473							26	139129						
27	134286							27	135977						

MOMENTS.

$$M_x = Rx - m.$$

SPAN, 93 feet.

x	R	m
1	176881	
2	173279	
3	169838	
4	166397	0 p_2
5	162956	
6	159516	
7	156075	
8	169838	
9	166397	
10	162957	125000 p_3
11	159516	
12	156075	
13	152795	
14	164354	
15	160913	
16	157472	320000 p_5
17	154031	
18	150752	
19	147472	
20	160913	
21	157472	
22	154031	
23	150752	645000 p_4
24	147472	
25	144192	
26	140913	
27	137634	

SPAN, 93 feet.

x	R	m
28	150752	
29	147472	
30	144192	
31	140913	
32	137634	1095000 p_5
33	134516	
34	131397	
35	128279	
36	125161	
37	144192	
38	140913	1900000 p_6
39	137634	
40	134516	
41	147472	
42	144192	2550000 p_7
43	140913	
44	137634	
45	150752	
46	147472	3275000 p_8
47	144192	

SPAN, 94 feet.

x	R	m
1	178563	
2	175000	
3	171436	
4	168031	0 p_2
5	164627	
6	161223	
7	157819	
8	171436	
9	168031	
10	164627	125000 p_3
11	161223	
12	157819	
13	154414	
14	166010	
15	162605	
16	159201	320000 p_5
17	155796	
18	152392	
19	149147	
20	162605	
21	159201	
22	155796	
23	152392	645000 p_4
24	149147	
25	145902	
26	142658	
27	139413	

SPAN, 94 feet.

x	R	m
28	152392	
29	149147	
30	145902	
31	142658	
32	139413	1095000 p_5
33	136168	
34	133083	
35	130000	
36	126913	
37	145902	
38	142658	1900000 p_6
39	139413	
40	136168	
41	149147	
42	145902	2550000 p_7
43	142658	
44	139413	
45	152392	
46	149147	3275000 p_8
47	145902	

MOMENTS.

$$M_x = Rx - m.$$

SPAN, 95 feet.

x	R	m	
1	180211		
2	176684		
3	173157	0	p_2
4	169631		
5	166263		
6	162894		
7	159526		
8	173157		
9	169631		
10	166263	125000	p_3
11	162894		
12	159526		
13	156157		
14	167631		
15	164262		
16	160894	320000	p_3
17	157525		
18	154157		
19	150788		
20	164262		
21	160894		
22	157525		
23	154157	640000	p_4
24	150788		
25	147578		
26	144367		
27	141157		
28	154157		
29	150788		
30	147578		
31	144367		
32	141157	1095000	p_5
33	137948		
34	134738		
35	131686		
36	128634		
37	147578		
38	144367	1900000	p_6
39	141157		
40	137948		

SPAN, 95 feet.

x	R	m	
41	150788		
42	147578	2550000	p_7
43	144367		
44	141157		
45	154157		
46	150788	3375000	p_8
47	147578		
48	144367		

SPAN, 96 feet.

x	R	m	
1	181822		
2	178333		
3	174843	0	p_2
4	171354		
5	167864		
6	164531		
7	161197		
8	174843		
9	171354		
10	167864	125000	p_3
11	164531		
12	161197		
13	157864		
14	169374		
15	165884		
16	162551	320000	p_3
17	159218		
18	155885		
19	152552		
20	165884		
21	162551		
22	159218		
23	155885	640000	p_4
24	152552		
25	149219		
26	146042		
27	142865		
28	155885		
29	152552		
30	149219		
31	146042		
32	142865	1095000	p_5
33	139689		
34	136512		
35	133335		
36	130314		
37	149219		
38	146042	1900000	p_6
39	142865		
40	139689		

SPAN, 96 feet.

x	R	m	
41	152552		
42	149219	2550000	p_7
43	146042		
44	142865		
45	155885		
46	152552	3375000	p_8
47	149219		
48	146042		

MOMENTS.

$$M_x = Rx - m.$$

Span, 97 feet.

x	R	m	
1	183525		
2	179948		
3	176494		
4	173041	0	p_1
5	169587		
6	166134		
7	162835		
8	176494		
9	173041		
10	169587	125000	p_2
11	166134		
12	162835		
13	159536		
14	171082		
15	167628		
16	164175	320000	p_3
17	160875		
18	157576		
19	154278		
20	167628		
21	164175		
22	160875		
23	157576	645000	p_4
24	154278		
25	150978		
26	147679		
27	144535		
28	157576		
29	154278		
30	150978		
31	147679		
32	144535	1095000	p_5
33	141391		
34	138246		
35	135102		
36	132958		
37	150978		
38	147679	1900000	p_6
39	144535		
40	141391		

Span, 97 feet.

x	R	m	
41	150978		
42	147679	2550000	p_7
43	144535		
44	141391		
45	157576		
46	154278		
47	150978	3275000	p_8
48	147679		
49	144535		

Span, 98 feet.

x	R	m	
1	185193		
2	181653		
3	178112		
4	174693	0	p_2
5	171275		
6	167857		
7	164438		
8	178112		
9	174693		
10	171275	125000	p_3
11	167857		
12	164438		
13	161173		
14	172754		
15	169335		
16	165917	320000	p_3
17	162500		
18	159234		
19	155968		
20	169335		
21	165917		
22	162500		
23	159234	645000	p_4
24	155968		
25	152703		
26	149438		
27	146173		
28	159234		
29	155968		
30	152703		
31	149438		
32	146173	1095000	p_5
33	143060		
34	139948		
35	136836		
36	133724		
37	152703		
38	149438	1900000	p_6
39	146173		
40	143060		

Span, 98 feet.

x	R	m	
41	155968		
42	152703	2550000	p_7
43	149438		
44	146173		
45	159234		
46	155968		
47	152703	3275000	p_8
48	149438		
49	146173		

MOMENTS.

$$M_x = Rx - m.$$

Span, 99 feet.

x	R	m	p
1	186828		
2	183323		
3	179818		
4	176313	0	p_2
5	172924		
6	169545		
7	166161		
8	179818		
9	176313		
10	172929	125000	p_3
11	169545		
12	166161		
13	162777		
14	174393		
15	171000		
16	167625	320000	p_3
17	164241		
18	160858		
19	157625		
20	171009		
21	167625		
22	164241		
23	160858	640000	p_4
24	157625		
25	154393		
26	151160		
27	147928		
28	160858		
29	157625		
30	154393		
31	151160		
32	147928	1090000	p_5
33	144696		
34	141615		
35	138534		
36	135454		
37	154393		
38	151160	1900000	p_6
39	147928		
40	144696		

Span, 99 feet.

x	R	m	p
41	157625		
42	154393	2590000	p_7
43	151160		
44	147928		
45	160858		
46	157625		
47	154393	3270000	p_8
48	151160		
49	147928		
50	144696		

Span, 100 feet.

x	R	m	p
1	188430		
2	184960		
3	181490		
4	178020	0	p_2
5	174550		
6	171200		
7	167850		
8	181490		
9	178020		
10	174550	125000	p_3
11	171200		
12	167850		
13	164500		
14	176000		
15	172650		
16	169300	320000	p_3
17	165950		
18	162600		
19	159250		
20	172650		
21	169300		
22	165950		
23	162600	640000	p_4
24	159250		
25	156050		
26	152850		
27	165950		
28	162600		
29	159250		
30	156050		
31	152850	1090000	p_5
32	149650		
33	146450		
34	143250		
35	140200		
36	137150		
37	156050		
38	152850	1900000	p_6
39	149650		
40	146450		

Span, 100 feet.

x	R	m	p
41	159250		
42	156050	2590000	p_7
43	152850		
44	149650		
45	162600		
46	159250		
47	156050	3270000	p_8
48	152850		
49	149650		
50	178020	4840000	p_{10}

MOMENTS.

$$M_x = Rx - m.$$

Span, 103 feet.

x	R	m	p
1	193281		
2	189796		
3	186310		
4	182941	0	p₂
5	179572		
6	176203		
7	172834		
8	186310		
9	182941		
10	179572	125000	p₃
11	176203		
12	172834		
13	169466		
14	181067		
15	177669		
16	174271	320000	p₃
17	170873		
18	167620		
19	164368		
20	177669		
21	174271		
22	170873		
23	167620	645000	p₄
24	164368		
25	161115		
26	157963		
27	154611		
28	167620		
29	164368		
30	161115		
31	157963		
32	154611	1095000	p₅
33	151504		
34	148398		
35	145290		
36	142183		
37	161115		
38	157963	1900000	p₆
39	154611		
40	151504		

Span, 103 feet.

x	R	m	p
41	164368		
42	161115		
43	157963	2550000	p₇
44	154611		
45	151504		
46	164368		
47	161115	3375000	p₈
48	157963		
49	154611		
50	182941		
51	179572	4840000	p₁₀
52	176203		

Span, 105 feet.

x	R	m	p
1	196438		
2	193019		
3	189600		
4	186180	0	p₂
5	182761		
6	179457		
7	176152		
8	189600		
9	186180		
10	182761	125000	p₃
11	179457		
12	176152		
13	172848		
14	184400		
15	180953		
16	177619	320000	p₃
17	173286		
18	170952		
19	167619		
20	180953		
21	177619		
22	173286		
23	170952	645000	p₄
24	167619		
25	164429		
26	161234		
27	158048		
28	170952		
29	167619		
30	164429		
31	161234	1095000	p₅
32	158048		
33	154858		
34	151668		
35	148621		
36	145574		
37	164429		
38	161234	1900000	p₆
39	158048		
40	154858		

Span, 105 feet.

x	R	m	p
41	167619		
42	164429		
43	161234	2550000	p₇
44	158048		
45	154858		
46	167619		
47	164429	3375000	p₈
48	161234		
49	158048		
50	170952	4075000	p₉
51	182761		
52	179457	4840000	p₁₀
53	176152		

MOMENTS.

$$M_x = Rx - m.$$

SPAN, 108 feet.			SPAN, 108 feet.			SPAN, 110 feet.			SPAN, 110 feet.		
x	R	m	x	R	m	x	R	m	x	R	m
1	200953	0 (p_2)	41	172684	2550000 (p_7)	1	203827	0 (p_2)	41	176017	2550000 (p_7)
2	197629		42	169443		2	200563		42	172726	
3	194305		43	166202		3	197300		43	169545	
4	190981		44	162962		4	194036		44	166362	
5	187657		45	159861		5	190772		45	163181	
6	184333					6	187509				
7	181009					7	184245				
8	194305	125000 (p_3)	46	172684	3375000 (p_8)	8	197300	125000 (p_3)	46	176017	3375000 (p_8)
9	190981		47	169443		9	194036		47	172726	
10	187657		48	166202		10	190772		48	169545	
11	184333		49	162962		11	187509		49	166362	
12	181009					12	184245				
13	177685					13	180981				
14	189333	320000 (p_3)	50	175925	4075000 (p_9)	14	192581	320000 (p_3)	50	179308	4075000 (p_9)
15	185981		51	172684		15	189181		51	176017	
16	182629					16	185890				
17	179277		52	184333	4840000 (p_{10})	17	182599		52	187509	4840000 (p_{10})
18	175925		53	181009		18	179308		53	184245	
19	172684		54	177685		19	176017		54	180981	
20	185981	645000 (p_4)				20	189181	645000 (p_4)			
21	182629					21	185890				
22	179277					22	182599		55	203827	6240000 (p_{11})
23	175925					23	179308				
24	172684					24	176017				
25	169443					25	172726				
26	166202					26	169545				
27	162962					27	166362				
28	175925	1095000 (p_5)				28	179308	1095000 (p_5)			
29	172684					29	176017				
30	169443					30	172726				
31	166202					31	169545				
32	162962					32	166362				
33	159861					33	163181				
34	156759					34	160000				
35	153657					35	156953				
36	150556					36	153908				
37	169443	1900000 (p_6)				37	172726	1900000 (p_6)			
38	166202					38	169545				
39	162962					39	166362				
40	159861					40	163181				

MOMENTS.

$$M_x = Rx - m.$$

Span, 113 feet.

x	R	m	p
1	208159		
2	204876		
3	201592		
4	198415	0	p_2
5	195238		
6	192061		
7	188884		
8	201592		
9	198415		
10	195238	125000	p_3
11	192061		
12	188884		
13	185707		
14	197397		
15	194088		
16	190778	320000	p_3
17	187470		
18	184160		
19	180957		
20	194088		
21	190778		
22	187470		
23	184160		
24	180957	649000	p_4
25	177753		
26	174550		
27	171346		
28	168144		
29	180957		
30	177753		
31	174550		
32	171346	1095000	p_5
33	168144		
34	165046		
35	161949		
36	158852		
37	177753		
38	174550	1900000	p_6
39	171346		
40	168144		

Span, 113 feet.

x	R	m	p
41	180957		
42	177753		
43	174550	2590000	p_7
44	171346		
45	168144		
46	180957		
47	177753	3275000	p_8
48	174550		
49	171346		
50	184160		
51	180957	4075000	p_9
52	177753		
53	174550		
54	185707	4840000	p_{10}
55	208159		
56	204876	6240000	p_{11}
57	201592		

Span, 115 feet.

x	R	m	p
1	210091		
2	207765		
3	204539		
4	201313	0	p_1
5	198086		
6	194965		
7	191843		
8	204539		
9	201313		
10	198086	125000	p_2
11	194965		
12	191843		
13	188721		
14	200468		
15	197217		
16	193964	320000	p_3
17	190712		
18	187460		
19	184207		
20	197217		
21	193964		
22	190712		
23	187460		
24	184207	649000	p_4
25	180956		
26	177809		
27	174660		
28	171512		
29	184207		
30	180956		
31	177809		
32	174660	1095000	p_5
33	171512		
34	168364		
35	165216		
36	162172		

Span, 115 feet.

x	R	m	p
37	180956		
38	177809		
39	174660	1900000	p_6
40	171512		
41	168364		
42	180956		
43	177809	2590000	p_7
44	174660		
45	171512		
46	184207		
47	180956		
48	177809	3275000	p_8
49	174660		
50	171512		
51	184207		
52	180956	4075000	p_9
53	177809		
54	174660		
55	200468	5690000	p_{10}
56	207765		
57	204539	6240000	p_{11}
58	201313		

MOMENTS.

$$M_x = Rx - m.$$

Span, 120 feet.			Span, 120 feet.			Span, 125 feet.			Span, 125 feet.		
x	R	m	x	R	m	x	R	m	x	R	m
1	218058		38	185883	1900000 p_6	1	225040		40	187424	1900000 p_6
2	214866		39	182766		2	221880		41	184432	
3	211675	0 p_1	40	179650		3	218720	0 p_2	42	181440	
4	208483		41	176533		4	215560		43	178448	
5	205291		42	189000	2550000 p_7	5	212400		44	190416	2550000 p_7
6	202200		43	185883		6	209336		45	187424	
7	199108		44	182766		7	206272		46	184432	
			45	179650		8	203208		47	181440	
8	211675		46	176533					48	178448	
9	208483		47	189000	3275000 p_8	9	215024	120000 p_2	49	190416	3275000 p_8
10	205291	125000 p_3	48	185883					50	187424	
11	202200		49	182766		10	212400		51	184432	
12	199108		50	179650		11	209336		52	181440	
13	196016		51	176533		12	206272	120000 p_3	53	178448	
						13	203208				
14	207800		52	189000	4075000 p_9				54	190416	4075000 p_9
15	204583		53	185883		14	215024		55	187424	
16	201466		54	182766		15	211840		56	184432	
17	198350	330000 p_2	55	179650		16	208752		57	181440	
18	195233		56	176533		17	205664	330000 p_3			
19	192116					18	202576		58	205664	5650000 p_{10}
20	189000		57	201466	5650000 p_{10}	19	199488		59	202576	
			58	198350		20	196400		60	199488	
21	201466					21	193408				
22	198350		59	205291	6240000 p_{11}				61	206272	6240000 p_{11}
23	195233		60	202200		22	205664		62	203208	
24	192116	64000 p_4				23	202576				
25	189000					24	199488		63	215024	7170000 p_{11}
26	185883					25	196400	64000 p_4			
27	182766					26	193408				
28	179650					27	190416				
						28	187424				
29	192116					29	184432				
30	189000					30	181440				
31	185883										
32	182766	1095000 p_5				31	193408				
33	179650					32	190416				
34	176533					33	187424				
35	173416					34	184432				
36	170389					35	181440	1095000 p_5			
37	167372					36	178448				
						37	175456				
						38	172464				
						39	169472				

MOMENTS.

$$M_x = Rx - m.$$

Span, 130 feet.			Span, 130 feet.			Span, 135 feet.			Span, 135 feet.		
x	R	m	x	R	m	x	R	m	x	R	m
1	231946		40	194783		1	238429		40	202042	
2	228815		41	191813		2	235414		41	199094	
3	225684		42	188844	1900000, p_6	3	232400		42	196146	
4	222553	0, p_2	43	185968		4	229385		43	193287	1900000, p_6
5	219422		44	183090		5	226370	0, p_2	44	190428	
6	216383					6	223354		45	187568	
7	213345		45	194783		7	220339				
8	210306		46	191813		8	217324		46	199094	
			47	188844	2550000, p_7				47	196146	
9	222152		48	185968		9	228498		48	193287	2550000, p_7
10	218998	120000, p_2	49	183090		10	226072	120000, p_2	49	190428	
11	215936					11	223035		50	187568	
			50	194783		12	219998				
			51	191813		13	216961		51	199094	
12	213345	125000, p_3	52	188844	3275000, p_8				52	196146	
13	210306		53	185968		14	228498		53	193287	3275000, p_8
			54	183090		15	226072		54	190428	
14	222152					16	223035		55	187568	
15	218998		55	194783		17	219998	330000, p_3			
16	215936		56	191813	4075000, p_9	18	216961		56	199094	
17	212875	320000, p_3	57	188844		19	213924		57	196146	4075000, p_9
18	209814		58	185968		20	210887		58	193287	
19	206752					21	207939		59	190428	
20	203690		59	209814							
21	200721		60	206752	5650000, p_{10}	22	219998		60	213924	
			61	203690		23	216961		61	210887	5650000, p_{10}
22	212875					24	213924		62	207939	
23	209814					25	210887		63	204991	
24	206752		62	210306	6240000, p_{11}	26	207939	645000, p_4			
25	203690					27	204991		64	226072	
26	200721	645000, p_4				28	202042		65	223035	
27	197752					29	199094		66	219998	7170000, f_{11}
28	194783		63	222152		30	196146		67	216961	
29	191813		64	218998	7170000, p_{11}				68	213924	
30	188844		65	215936		31	207939				
						32	204991				
31	200721					33	202042				
32	197752					34	199094				
33	194783					35	196146	1095000, p_5			
34	191813					36	193287				
35	188844	1095000, p_5				37	190428				
36	185968					38	187568				
37	183090					39	184708				
38	180214										
39	177326										

MOMENTS.

$$M_x = Rx - m.$$

SPAN, 140 feet.

x	R	m
1	244792	
2	241800	
3	238807	0
4	235814	p_1
5	232821	
6	229914	
7	227007	
8	224100	
9	236084	
10	233070	120000
11	230056	p_2
12	227042	
13	224028	
14	236084	
15	233070	
16	230056	
17	227042	320000
18	224028	p_3
19	221014	
20	218000	
21	215070	
22	227042	
23	224028	
24	221014	
25	218000	64,000
26	215070	p_4
27	212142	
28	209213	
29	206285	
30	203356	
31	215070	
32	212142	
33	209213	
34	206285	
35	203356	109,000
36	200514	p_5
37	197671	
38	194827	
39	191985	
40	189142	

SPAN, 140 feet.

x	R	m
41	206285	
42	203356	1900000
43	200514	p_6
44	197671	
45	194827	
46	206285	
47	203356	
48	200514	2550000
49	197671	p_7
50	194827	
51	206285	
52	203356	
53	200514	3275000
54	197671	p_8
55	194827	
56	206285	
57	203356	4075000
58	200514	p_9
59	197671	
60	194827	
61	218000	5650000
62	215070	p_{10}
63	212142	
64	233070	
65	230056	
66	227042	
67	224028	7170000
68	221014	p_{11}
69	218000	
70	215070	

SPAN, 145 feet.

x	R	m
1	251131	
2	248158	
3	245186	0
4	242213	p_1
5	239241	
6	236351	
7	233461	
8	230571	
9	242578	
10	239585	120000
11	236674	p_2
12	233764	
13	230853	
14	242578	
15	239585	
16	236674	
17	233764	320000
18	230853	p_3
19	227943	
20	225032	
21	222122	
22	219211	
23	230853	
24	227943	
25	225032	
26	222122	64,000
27	219211	p_4
28	216301	
29	213391	
30	210481	
31	207653	
32	219211	
33	216301	
34	213391	
35	210481	109,000
36	207653	p_5
37	204825	
38	201998	
39	199170	
40	196343	

SPAN, 145 feet.

x	R	m
41	213391	
42	210481	1900000
43	207653	p_6
44	204825	
45	201998	
46	213391	
47	210481	
48	207653	2550000
49	204825	p_7
50	201998	
51	213391	
52	210481	
53	207653	3275000
54	204825	p_8
55	201998	
56	199170	
57	210481	4075000
58	207653	p_9
59	204825	
60	201998	
61	225032	
62	222122	5650000
63	219211	p_{10}
64	216301	
65	213391	
66	233764	
67	230853	
68	227943	
69	225032	7170000
70	222122	p_{11}
71	219211	
72	216301	
73	213391	

MOMENTS.

$$M_x = Rx - m.$$

Span, 150 feet.

x	R	m	
1	257446		
2	254493		
3	251540		
4	248586	0	p_2
5	245633		
6	242760		
7	239866		
8	237013		
9	249040		
10	246066		
11	243173	120000	p_3
12	240280		
•13	237386		
14	249040		
15	246066		
16	243173		
17	240280		
18	237386	320000	p_3
19	234493		
20	231600		
21	228786		
22	225973		
23	237386		
24	234493		
25	231600		
26	228786		
27	225973	64500	p_4
28	223160		
29	220346		
30	217533		
31	214720		
32	225973		
33	223160		
34	220346		
35	217533		
36	214720	105000	p_5
37	211906		
38	209093		
39	206280		
40	203466		
41	200733		

Span, 150 feet.

x	R	m	
42	217533		
43	214720		
44	211906	1900000	p_6
45	209093		
46	206280		
47	217533		
48	214720		
49	211906	2550000	p_7
50	209093		
51	206280		
52	217533		
53	214720		
54	211906	327500	p_8
55	209093		
56	206280		
57	217533		
58	214720		
59	211906	407500	p_9
60	209093		
61	206280		
62	203466		
63	225973		
64	223160		
65	220346	565000	p_{10}
66	217533		
67	214720		
68	234493		
69	231600		
70	228786		
71	225973		
72	223160	7170000	p_{11}
73	220346		
74	217533		
75	214720		

Span, 155 feet.

x	R	m	
1	263741		
2	260806		
3	257870		
4	254935	0	p_2
5	252000		
6	249141		
7	246283		
8	243425		
9	255470		
10	252515		
11	249638	120000	p_3
12	246761		
13	243883		
14	255470		
15	252515		
16	249638		
17	246761		
18	243883	320000	p_3
19	241006		
20	238128		
21	235328		
22	232528		
23	243883		
24	241006		
25	238128		
26	235328		
27	232528	64500	p_4
28	229727		
29	226926		
30	224126		
31	221403		
32	218681		
33	229727		
34	226926		
35	224126		
36	221403		
37	218681	105000	p_5
38	215958		
39	213236		
40	210515		
41	207793		
42	205070		

Span, 155 feet.

x	R	m	
43	221403		
44	218681		
45	215958	1900000	p_6
46	213236		
47	210515		
48	221403		
49	218681		
50	215958	2550000	p_7
51	213236		
52	210515		
53	207793		
54	218681		
55	215958		
56	213236	327500	p_8
57	210515		
58	207793		
59	218681		
60	215958		
61	213236	407500	p_9
62	210515		
63	207793		
64	229727		
65	226926		
66	224126	565000	p_{10}
67	221403		
68	218681		
69	238128		
70	235328		
71	232528		
72	229727		
73	226926	7170000	p_{11}
74	224126		
75	221403		
76	218681		
77	215958		
78	213236		

MOMENTS.

$$M_x = Rx - m.$$

SPAN, 160 feet.

x	R	m
1	269725	0 p_1
2	266881	
3	264037	
4	261193	
5	258350	
6	255506	
7	252662	
8	249818	
9	261880	120000 p_2
10	258942	
11	256079	
12	253197	
13	250354	
14	261880	320000 p_3
15	258942	
16	256079	
17	253197	
18	250354	
19	247491	
20	244628	
21	241840	
22	239053	
23	236255	
24	247491	645000 p_4
25	244628	
26	241840	
27	239053	
28	236255	
29	233467	
30	230680	
31	227967	
32	225254	
33	236255	1095000 p_5
34	233467	
35	230680	
36	227967	
37	225254	
38	222541	
39	219829	
40	217116	
41	214486	
42	211849	
43	227967	1900000 p_6
44	225254	
45	222541	
46	219829	
47	217116	
48	214486	
49	225254	2550000 p_7
50	222541	
51	219829	
52	217116	
53	214486	
54	211849	
55	222541	3275000 p_8
56	219829	
57	217116	
58	214486	
59	211849	
60	209212	
61	219829	4075000 p_9
62	217116	
63	214486	
64	211849	
65	233467	965000 p_{10}
66	230680	
67	227967	
68	225254	
69	244628	7170000 p_{11}
70	241840	
71	239053	
72	236255	
73	233467	
74	230680	
75	227967	
76	225254	
77	222541	
78	219829	
79	217116	
80	227967	8245000 p_{12}

SPAN, 165 feet.

x	R	m
1	275624	0 p_1
2	272793	
3	269963	
4	267133	
5	264303	
6	261545	
7	258787	
8	256030	
9	268181	120000 p_2
10	265333	
11	262484	
12	259636	
13	256787	
14	253938	
15	265333	320000 p_3
16	262484	
17	259636	
18	256787	
19	253938	
20	251089	
21	248313	
22	245537	
23	242761	
24	253938	645000 p_4
25	251089	
26	248313	
27	245537	
28	242761	
29	239986	
30	237211	
31	234507	
32	231804	
33	242761	1095000 p_5
34	239986	
35	237211	
36	234507	
37	231804	
38	229101	

MOMENTS.

$$M_x = Rx - m.$$

Span, 165 feet.

x	R	m
39	226397	
40	223696	
41	221072	1095000 p_5
42	218447	
43	215823	
44	231804	
45	229101	
46	226397	1900000 p_6
47	223696	
48	221072	
49	231804	
50	229101	
51	226397	2550000 p_7
52	223696	
53	221072	
54	218447	
55	229101	
56	226397	
57	223696	3275000 p_8
58	221072	
59	218447	
60	215823	
61	226397	
62	223696	
63	221072	4075000 p_9
64	218447	
65	215823	
66	237211	
67	234507	5650000 p_{10}
68	231804	
69	229101	
70	248313	
71	245537	
72	242761	7170000 p_{11}
73	239986	
74	237211	
75	234507	

Span, 165 feet.

x	R	m
76	231804	
77	229101	
78	226397	7170000 p_{11}
79	223696	
80	221072	
81	231804	
82	229101	8245000 p_{12}
83	226397	

Span, 170 feet.

x	R	m
1	281535	
2	278717	
3	275900	
4	273082	0 p_1
5	270264	
6	267517	
7	264770	
8	262023	
9	274188	
10	271353	
11	268687	
12	265823	120000 p_2
13	263059	
14	260294	
15	271353	
16	268687	
17	265823	
18	263059	
19	260294	320000 p_3
20	257529	
21	254764	
22	252000	
23	249235	
24	260294	
25	257529	
26	254764	
27	252000	
28	249235	645000 p_4
29	246470	
30	243705	
31	241010	
32	238317	
33	235624	
34	246470	
35	243705	
36	241010	
37	238317	1095000 p_5
38	235624	
39	232929	

Span, 170 feet.

x	R	m
40	230235	
41	227611	
42	224987	1095000 p_5
43	222364	
44	238317	
45	235624	
46	232929	1900000 p_6
47	230235	
48	227611	
49	238317	
50	235624	
51	232929	2550000 p_7
52	230235	
53	227611	
54	224987	
55	235624	
56	232929	
57	230235	3275000 p_8
58	227611	
59	224987	
60	222364	
61	232929	
62	230235	
63	227611	4075000 p_9
64	224987	
65	222364	
66	243705	
67	241010	5650000 p_{10}
68	238317	
69	235624	
70	254764	
71	252000	
72	249235	
73	246470	7170000 p_{11}
74	243705	
75	241010	
76	238317	

MOMENTS.

$$M_x = Rx - m.$$

SPAN, 170 feet.

x	R	m
77	235624	
78	232930	7170000 p_{11}
79	230235	
80	227611	
81	238317	
82	235624	
83	232929	8245000 p_{12}
84	230235	
85	227611	

SPAN, 175 feet.

x	R	m
1	287451	
2	284645	
3	281840	0 p_2
4	279034	
5	276228	
6	273491	
7	270754	
8	268017	
9	280194	
10	277370	
11	274616	120000 p_3
12	271862	
13	269108	
14	266353	
15	277370	
16	274616	
17	271862	
18	269108	
19	266353	330000 p_3
20	263600	
21	260914	
22	258228	
23	255542	
24	252857	
25	263600	
26	260914	
27	258228	
28	255542	
29	252857	645000 p_4
30	250171	
31	247485	
32	244800	
33	242114	
34	252857	
35	250171	
36	247485	
37	244800	1095000 p_5
38	242114	
39	239428	
40	236742	
41	234125	

SPAN, 175 feet.

x	R	m
42	231508	1095000 p_5
43	228890	
44	244800	
45	242114	
46	239428	1900000 p_6
47	236742	
48	234125	
49	244800	
50	242114	
51	239428	2550000 p_7
52	236742	
53	234125	
54	244800	
55	242114	
56	239428	
57	236742	3275000 p_8
58	234125	
59	231508	
60	228890	
61	239428	
62	236742	
63	234125	4075000 p_9
64	231508	
65	228890	
66	250171	
67	247485	
68	244800	5650000 p_{10}
69	242114	
70	239428	
71	258228	
72	255542	
73	252857	7170000 p_{11}
74	250171	
75	247485	
76	244800	

SPAN, 175 feet.

x	R	m
77	242114	
78	239428	7170000 p_{11}
79	236742	
80	234125	
81	244800	
82	242114	
83	239428	
84	236742	8245000 p_{12}
85	234125	
86	231508	
87	228890	
88	226273	

APPENDIX.

MOMENTS.

$$M_x = Rx - m.$$

Span, 180 feet (x = 1 to 90)

x	R	m	p
1	293372	0	p_1
2	290577		
3	287783		
4	284988		
5	282194		
6	279466		
7	276738		
8	274010		
9	286200	120000	p_2
10	283388		
11	286644		
12	277900		
13	275155		
14	272410		
15	283388	320000	p_3
16	280644		
17	277900		
18	275155		
19	272410		
20	269665		
21	266988		
22	264310		
23	261633		
24	258955		
25	269665	645000	p_4
26	266988		
27	264310		
28	261633		
29	258955		
30	256277		
31	253655		
32	251055		
33	248444		
34	245833		
35	256277	1095000	p_5
36	253655		
37	251055		
38	248444		
39	245833		
40	243221		
41	240610		
42	238000		
43	235391		
44	232777		
45	248444	1900000	p_6
46	245833		
47	243221		
48	240610		
49	238000		
50	248444	2550000	p_7
51	245833		
52	243221		
53	240610		
54	238000		
55	235391		
56	245833	3275000	p_8
57	243221		
58	240610		
59	238000		
60	235391		
61	245833	4075000	p_9
62	243221		
63	240610		
64	238000		
65	235391		
66	232777		
67	230166		
68	251055	5650000	p_{10}
69	248444		
70	245833		
71	243221		
72	240610		
73	258955	7170000	p_{11}
74	256277		
75	253655		
76	251055		
77	248444		
78	245833		
79	243221		
80	240610		
81	238000		
82	235391		
83	245833	8245000	p_{12}
84	243221		
85	240610		
86	238000		
87	235391		
88	232777		
89	230166		
90	227621		

For all values of x below x = 67, a uniformly distributed load of 3000 lbs. per ft. will give greater moments than this table.

Span, 185 feet (x = 1 to 40)

x	R	m	p
1	299037	0	p_1
2	296318		
3	293600		
4	290881		
5	288162		
6	285443		
7	282724		
8	280005		
9	292204	120000	p_2
10	289404		
11	286669		
12	283934		
13	281120		
14	278464		
15	289404	320000	p_3
16	286669		
17	283934		
18	281120		
19	278464		
20	275729		
21	273058		
22	270388		
23	267717		
24	265047		
25	275729	645000	p_4
26	273058		
27	270388		
28	267717		
29	265047		
30	262377		
31	259770		
32	257165		
33	254560		
34	251955		
35	262377	1095000	p_5
36	259770		
37	257165		
38	254560		
39	251955		
40	249350		

MOMENTS.

$$M_x = Rx - m.$$

Span, 185 feet.

x	R	m	p
41	246810		
42	244269	1095000	p_5
43	241729		
44	239188		
45	254560		
46	251955		
47	249350		
48	246810	1900000	p_6
49	244269		
50	241729		
51	239188		
52	249350		
53	246810		
54	244269	2550000	p_7
55	241729		
56	239188		
57	236647		
58	246810		
59	244269	3275000	p_8
60	241729		
61	239188		
62	236647		
63	246810		
64	244269		
65	241729	4075000	p_9
66	239188		
67	236647		
68	234106		
69	254560		
70	251955		
71	249350	5650000	p_{10}
72	246810		
73	244269		
74	262377		
75	259770		
76	257165	7170000	p_{11}
77	254560		
78	251955		
79	249350		

Span, 185 feet.

x	R	m	p
80	246810		
81	244269		
82	241729	7170000	p_{11}
83	239188		
84	236647		
85	246810		
86	244269		
87	241729		
88	239188		
89	236647	8245000	p_{12}
90	234106		
91	231566		
92	229024		
93	230284		

For all values of x below x = 57, a uniformly distributed load of 3000 lbs. per ft. will give greater moments than this table.

Span, 190 feet.

x	R	m	p
1	304657		
2	301947		
3	299236		
4	296526		
5	293815	0	p_2
6	291167		
7	288520		
8	285872		
9	298145		
10	295418		
11	292692	120000	p_3
12	289965		
13	287238		
14	284511		
15	295418		
16	292692		
17	289965		
18	287238		
19	284511		
20	281785	320000	p_1
21	279122		
22	276459		
23	273795		
24	271132		
25	281785		
26	279122		
27	276459		
28	273795		
29	271132		
30	268469	645000	p_4
31	265869		
32	263269		
33	260669		
34	258069		
35	268469		
36	265869		
37	263269		
38	260669	1095000	p_5
39	258069		
40	255469		

Span, 190 feet.

x	R	m	p
41	252932		
42	250396		
43	247859	1095000	p_5
44	245322		
45	242785		
46	258069		
47	255469		
48	252932	1900000	p_6
49	250396		
50	247859		
51	258069		
52	255469		
53	252932		
54	250396	2550000	p_7
55	247859		
56	245322		
57	242785		
58	252932		
59	250396		
60	247859	3275000	p_8
61	245322		
62	242785		
63	240315		
64	250396		
65	247859		
66	245322	4075000	p_9
67	242785		
68	240315		
69	260669		
70	258069		
71	255469	5650000	p_{10}
72	252932		
73	250396		
74	263469		
75	265869		
76	263269	7170000	p_{11}
77	260669		
78	258069		

MOMENTS.

$$M_x = Rx - m.$$

SPAN, 190 feet.			SPAN, 195 feet.			SPAN, 195 feet.			SPAN, 195 feet.		
x	R	m	x	R	m	x	R	m	x	R	m
79	255469		1	310297		41	259056		79	261589	
80	252932		2	307595		42	256522		80	259056	
81	250396	7170000	3	304892		43	253989	1095000	81	256522	
82	247859	f_{11}	4	302189	0	44	251455	f_3	82	253989	7170000
83	245322		5	299487	f_1	45	248922		83	251455	f_{11}
84	242785		6	296846					84	248922	
85	240315		7	294205		46	264184		85	246450	
			8	291564		47	261589				
86	250396					48	259056	1900000	86	256522	
87	247859		9	303846		49	256522	f_6	87	253989	
88	245322		10	301128		50	253989		88	251455	
89	242785		11	298471		51	251455		89	248922	
90	240315	8245000	12	295815	130000				90	246450	
91	237841	f_{12}	13	293158	f_2	52	261589		91	243978	
92	235368		14	290502		53	259056		92	241506	8245000
93	232894		15	287845		54	256522	2550000	93	239035	f_{12}
94	230420					55	253989	f_7	94	236564	
95	227946		16	298471		56	251455		95	234153	
			17	295815		57	248922		96	231743	
			18	293158					97	229333	
			19	290502		58	259056		98	226923	
			20	290502	320000	59	256522				
			21	285189	f_3	60	253989	3175000			
			22	282532		61	251455	f_8			
			23	279876		62	248922				
			24	277219							
						63	259056				
			25	287845		64	256522				
			26	285189		65	253989	4075000			
			27	282532		66	251455	f_9			
			28	279876		67	248922				
			29	277219	645000	68	246450				
			30	274563	f_4						
			31	271969		69	266779				
			32	269374		70	264184				
			33	266779		71	261589	9650000			
			34	264184		72	259056	f_{10}			
						73	256522				
			35	274563							
			36	271969		74	274563				
			37	269374	1095000	75	271964	7170000			
			38	266779	f_5	76	269374	f_{11}			
			39	264184		77	266779				
			40	261589		78	264184				

For all values of x below $x = 50$, a uniformly distributed load of 3000 lbs. per ft. will give greater moments than this table.

For all values of x below $x = 45$, a uniformly distributed load of 3000 lbs. per ft. gives greater moments than this table.

MOMENTS.

$$M_x = Rx - m.$$

	Span, 200 feet.			Span, 200 feet.			Span, 200 feet.			Span, 210 feet.	
x	R	m	x	R	m	x	R	m	x	R	m
1	315955	0 (p_2)	41	265170	1095000 (p_5)	79	267700	7170000 (p_{11})	1	327090	0 (p_2)
2	313260		42	262640		80	265170		2	324466	
3	310565		43	260110		81	262640		3	321842	
4	307870		44	257580		82	260110		4	319219	
5	305175		45	255050		83	257580		5	316595	
6	302540		46	270290	1900000 (p_6)	84	255050		6	313971	
7	299905		47	267700		85	252580		7	311347	
8	297270		48	265170		86	262640	8245000 (p_{12})	8	308723	
9	309560	120000 (p_3)	49	262640		87	260110		9	321028	100000 (p_3)
10	306850		50	260110		88	257580		10	318333	
11	304200		51	257580		89	255050		11	315694	
12	301550		52	267700	2550000 (p_7)	90	252580		12	313056	
13	298900		53	265170		91	250110		13	310418	
14	296250		54	262640		92	247640		14	307780	
15	293600		55	260110		93	245170		15	305142	
16	291010		56	257580		94	242700		16	302561	
17	301550	320000 (p_1)	57	255050		95	240290		17	313056	300000 (p_3)
18	298900		58	265170	3273000 (p_8)	96	237880		18	310418	
19	296250		59	262640		97	235470		19	307780	
20	293600		60	260110		98	233060		20	305142	
21	291010		61	257580		99	230650		21	302561	
22	288420		62	255050		100	228300		22	299980	
23	285750		63	252580					23	297400	
24	283240		64	262640	4079000 (p_9)				24	294818	
25	280650		65	260110					25	292237	
26	291010	640000 (p_4)	66	257580					26	302561	640000 (p_4)
27	288420		67	255050					27	299980	
28	285750		68	252580					28	297400	
29	283240		69	250110					29	294818	
30	280650		70	270290	5650000 (p_{10})				30	292237	
31	278060		71	267700					31	289713	
32	275470		72	265170					32	287190	
33	272880		73	262640					33	284666	
34	270290		74	260110					34	282142	
35	280650	1095000 (p_5)	75	278060	7170000 (p_{11})				35	279618	
36	278060		76	275470					36	289713	1095000 (p_5)
37	275470		77	272880					37	287190	
38	272880		78	270290					38	284666	
39	270290								39	282142	
40	267700								40	279618	

For all values of x below x = 41, a uniformly distributed load of 3000 lbs. per ft. will give greater moments than this table.

MOMENTS.

$$M_x = Rx - m.$$

SPAN, 210 feet.

x	R	m	
41	277152		
42	274684		
43	272218	1095000	p_3
44	269751		
45	267284		
46	264818		
47	279618		
48	277152		
49	274684	1900000	p_6
50	272218		
51	269751		
52	267284		
53	277152		
54	274684		
55	272218	2550000	p_7
56	269751		
57	267284		
58	264818		
59	274684		
60	272218		
61	269751	3275000	p_8
62	267284		
63	264818		
64	262351		
65	272218		
66	269751		
67	267284	4075000	p_9
68	264818		
69	262351		
70	259884		
71	279618		
72	277152		
73	274684	9565000	p_{10}
74	272218		
75	269751		
76	267284		

SPAN, 210 feet.

x	R	m	
77	284666		
78	282142		
79	279618		
80	277152		
81	274684		
82	272218	7170000	p_{11}
83	269751		
84	267284		
85	264818		
86	262351		
87	259884		
83	269751		
89	267284		
90	264818		
91	262351		
92	259884		
93	257417		
94	254951	8245000	p_{12}
95	252542		
96	250132		
97	247723		
98	245313		
99	242904		
100	240551		
101	250132		
102	247723		
103	245313	9445000	p_{13}
104	242904		
105	240551		

For all values of x below x = 33, a uniformly distributed load of 3000 lbs. per ft. will give greater moments than this table.

SPAN, 220 feet.

x	R	m	
1	337977		
2	335363		
3	332750		
4	330136	0	p_1
5	327522		
6	324963		
7	322404		
8	319845		
9	332217		
10	329590		
11	327017		
12	324445	120000	p_2
13	321872		
14	319300		
15	316726		
16	327017		
17	324445		
18	321872		
19	319300		
20	316726	320000	p_3
21	314154		
22	311581		
23	309008		
24	306436		
25	303862		
26	314154		
27	311581		
28	309008		
29	306436		
30	303862	649000	p_4
31	301344		
32	298826		
33	296308		
34	293790		
35	291271		
36	301344		
37	298826		
38	296308	1095000	p_5
39	293790		
40	291271		

SPAN, 220 feet.

x	R	m	
41	288808		
42	286344		
43	283881	1095000	p_3
44	281418		
45	278954		
46	276544		
47	291271		
48	288808		
49	286344	1900000	p_6
50	283881		
51	281418		
52	278954		
53	276544		
54	286344		
55	283881		
56	281418	2550000	p_7
57	278954		
58	276544		
59	274135		
60	283881		
61	281418		
62	278954	3275000	p_8
63	276544		
64	274135		
65	271726		
66	269317		
67	278954		
68	276544		
69	274135	4075000	p_9
70	271726		
71	269317		
72	288808		
73	286344		
74	283881	9565000	p_{10}
75	281418		
76	278954		

MOMENTS.

$$M_x = Rx - m.$$

Span, 220 feet.

x	R	m
77	296308	
78	293790	
79	291271	
80	288808	
81	286344	7170000 p_{11}
82	283881	
83	281418	
84	278954	
85	276544	
86	274135	
87	271726	
88	269317	
89	266909	
90	276544	
91	274135	
92	271726	
93	269317	
94	266909	
95	264553	8245000 p_{12}
96	262200	
97	259844	
98	257490	
99	255136	
100	252781	
101	248072	
102	259844	
103	257490	9445000 p_{13}
104	255136	
105	252781	
106	262200	
107	259844	
108	257490	10770000 p_{14}
109	255136	
110	252781	

For all values of x below x = 93, a uniformly distributed load of 3000 lbs. per ft. will give greater moments than this table.

Span, 230 feet.

x	R	m
1	348960	
2	346356	
3	343752	
4	341147	0 p_1
5	338543	
6	335991	
7	333440	
8	330886	
9	343268	
10	340651	
11	338086	120000 p_2
12	335521	
13	332955	
14	330390	
15	327825	
16	338086	
17	335521	
18	332955	
19	330390	
20	327825	320000 p_3
21	325312	
22	322800	
23	320286	
24	317773	
25	315260	
26	325312	
27	322800	
28	320286	
29	317773	
30	315260	
31	312800	645000 p_4
32	310338	
33	307877	
34	305416	
35	302956	
36	300495	
37	310338	
38	307877	1093000 p_5
39	305416	
40	302956	

Span, 230 feet.

x	R	m
41	300495	
42	298033	
43	295573	
44	293112	1095000 p_5
45	290651	
46	288243	
47	285834	
48	300495	
49	298033	
50	295573	
51	293112	1900000 p_6
52	290651	
53	288243	
54	298033	
55	295573	
56	293112	
57	290651	2550000 p_7
58	288243	
59	285834	
60	295573	
61	293112	
62	290651	
63	288243	3275000 p_8
64	285834	
65	283425	
66	281017	
67	290651	
68	288243	
69	285834	4075000 p_9
70	283425	
71	281017	
72	300495	
73	298033	
74	295573	
75	293112	3650000 p_{10}
76	290651	
77	288243	

Span, 330 feet.

x	R	m
78	305416	
79	302956	
80	300495	
81	298033	
82	295573	
83	293112	7170000 p_{11}
84	290651	
85	288243	
86	285834	
87	283425	
88	281017	
89	278608	
90	288243	
91	285834	
92	283425	
93	281017	
94	278608	
95	276251	8245000 p_{12}
96	273894	
97	271538	
98	269182	
99	266826	
100	264520	
101	262216	
102	259912	
103	257608	
104	255303	
105	264520	
106	262216	9445000 p_{13}
107	259912	
108	257608	
109	266826	
110	264520	
111	262216	
112	259912	10770000 p_{14}
113	257608	
114	255303	
115	253051	

Below x = 6, a uniform load of 3000 lbs. per ft. gives greater moments than this table.

MOMENTS.

$$M_x = Rx - m.$$

Span, 240 feet.

x	R	m	p
1	359579	0	p₁
2	357033		
3	354487		
4	351941		
5	349395		
6	346900		
7	344404		
8	341908		
9	354350	120000	p₂
10	351791		
11	349232		
12	346675		
13	344116		
14	341558		
15	339000		
16	349232	320000	p₃
17	346675		
18	344116		
19	341558		
20	339000		
21	336491		
22	333982		
23	331475		
24	328966		
25	326458		
26	324000		
27	333982	640000	p₄
28	331475		
29	328966		
30	326458		
31	324000		
32	321541		
33	319082		
34	316625		
35	314166		
36	311798		
37	321541	1095000	p₅
38	319082		
39	316625		
40	314166		

Span, 240 feet.

x	R	m	p
41	311758	1095000	p₅
42	309350		
43	306941		
44	304532		
45	302125		
46	299766		
47	297408		
48	295050		
49	309350	1590000	p₆
50	306941		
51	304532		
52	302125		
53	299766		
54	297408		
55	306941	2550000	p₇
56	304532		
57	302125		
58	299766		
59	297408		
60	295050		
61	292691		
62	302125	3175000	p₈
63	299766		
64	297408		
65	295050		
66	292691		
67	290332		
68	299766	4075000	p₉
69	297408		
70	295050		
71	292691		
72	290332		
73	287975		
74	306941	5650000	p₁₀
75	304532		
76	302125		
77	299766		
78	297408		
79	295050		

Span, 240 feet.

x	R	m	p
80	311758	7170000	p₁₁
81	309350		
82	306941		
83	304532		
84	302125		
85	299766		
86	297408		
87	295050		
88	292691		
89	290332		
90	287975		
91	285616		
92	295050	8245000	p₁₂
93	292691		
94	290332		
95	287975		
96	285616		
97	283258		
98	280900		
99	278541		
100	276232		
101	273925		
102	271616		
103	269308		
104	278541	9445000	p₁₃
105	276232		
106	273925		
107	271616		
108	269308		
109	267000		
110	264741		
111	262482		
112	260225		
113	257966		
114	255708		
115	253500		
116	251291		
117	249082		
118	246875		
119	244666		
120	242509		

A uniform load of 3000 lbs. per ft. gives greater moments than this table.

Span, 250 feet.

x	R	m	p
1	370260	0	p₁
2	367720		
3	365180		
4	362640		
5	360100		
6	357608		
7	355166		
8	352624		
9	365072	120000	p₂
10	362520		
11	360016		
12	357512		
13	355008		
14	352504		
15	350000		
16	347544		
17	357512	320000	p₃
18	355008		
19	352504		
20	350000		
21	347544		
22	345088		
23	342632		
24	340176		
25	337720		
26	335264		
27	345088	640000	p₄
28	342632		
29	340176		
30	337720		
31	335264		
32	332808		
33	330352		
34	327896		
35	325440		
36	323032		
37	332808	1095000	p₅
38	330352		
39	327896		
40	325440		

MOMENTS.

$$M_x = Rx - m.$$

SPAN, 250 feet.

x	R	m	
41	323032		
42	320624		
43	318216		
44	315808	1095000	p5
45	313400		
46	311040		
47	308680		
48	306320		
49	320624		
50	318216		
51	315808	1900000	p6
52	313400		
53	311040		
54	308680		
55	318216		
56	315808		
57	313400		
58	311040	2550000	p7
59	308680		
60	306320		
61	303960		
62	313400		
63	311040		
64	308680		
65	306320	3775000	p8
66	303960		
67	301600		
68	299238		
69	308680		
70	306320		
71	303960		
72	301600	4075000	p9
73	299288		
74	296976		
75	315808		
76	313400		
77	311040	5690000	p10
78	308680		
79	306320		

SPAN, 250 feet.

x	R	m	
80	323032		
81	320624		
82	318216		
83	315808		
84	313400		
85	311040		
86	308680	7170000	p11
87	306320		
88	303960		
89	301600		
90	299288		
91	296976		
92	294664		
93	303960		
94	301600		
95	299288		
96	296076		
97	294664		
98	292352		
99	290040		
100	287776	8245000	p12
101	285512		
102	283248		
103	280984		
104	278720		
105	277456		
106	274192		
107	283248		
108	280984		
109	278720		
110	277456		
111	274192		
112	271928		
113	269664		
114	267400	9445000	p13
115	265184		
116	262968		
117	260752		
118	258536		
119	256320		
120	254152		

SPAN, 250 feet.

x	R	m	
121	262968		
122	260752		
123	258536	10770000	p14
124	256320		
125	254152		

For all values of x below x = 120, a uniformly distributed load of 2700 lbs. per ft. will give greater moments than this table.

SPAN, 260 feet.

x	R	m	
1	380857		
2	378369		
3	375880		
4	373392	0	p2
5	370903		
6	368415		
7	365926		
8	363438		
9	375892		
10	373345		
11	370845		
12	368345	120000	p8
13	365845		
14	363345		
15	360845		
16	358391		
17	368345		
18	365845		
19	363345		
20	360845		
21	358391		
22	355937	320000	p3
23	353484		
24	351030		
25	348576		
26	346169		
27	355937		
28	353484		
29	351030		
30	348576		
31	346169		
32	343761	64000	p4
33	341353		
34	338945		
35	336538		
36	334176		
37	331814		

MOMENTS.

$$M_x = Rx - m.$$

Span, 260 feet.

x	R	m	
38	341353		
39	338945		
40	336538		
41	334176		
42	331814	1095000	p_5
43	329453		
44	327091		
45	324730		
46	322369		
47	320007		
48	317645		
49	331814		
50	329453		
51	327091		
52	324730	1900000	p_6
53	322369		
54	320007		
55	317645		
56	327091		
57	324730		
58	322369	2550000	p_7
59	320007		
60	317645		
61	315284		
62	324730		
63	322369		
64	320007		
65	317645	3270000	p_8
66	315284		
67	312922		
68	310607		
69	320007		
70	317645		
71	315284	4075000	p_9
72	312922		
73	310607		
74	308291		

Span, 260 feet.

x	R	m	
75	327091		
76	324730		
77	322369	5650000	p_{10}
78	320007		
79	317645		
80	315284		
81	331814		
82	329453		
83	327091		
84	324730		
85	322369		
86	320007	7170000	p_{11}
87	317645		
88	315284		
89	312922		
90	310607		
91	308291		
92	305976		
93	315284		
94	312922		
95	310607		
96	308291		
97	305976		
98	303661		
99	301345		
100	299076	8245000	p_{12}
101	296807		
102	294537		
103	292269		
104	290000		
105	287776		
106	285553		
107	283330		
108	292269		
109	290000		
110	287776		
111	285553	9445000	p_{13}
112	283330		
113	281107		
114	278883		
115	276707		

Span, 260 feet.

x	R	m	
116	274530		
117	272353		
118	270176	9445000	p_{13}
119	268000		
120	268822		
121	263646		
122	272353		
123	270176		
124	268000		
125	268822		
126	263646	10770000	p_{14}
127	261468		
128	259291		
129	257115		
130	254984		

For all values of x below x = 105, a uniformly distributed load of 1700 lbs. per ft. will give greater moments than this table.

Span, 270 feet.

x	R	m	
1	391292		
2	388807		
3	386322		
4	383837	0	p_2
5	381351		
6	378911		
7	376470		
8	374029		
9	386532		
10	384036		
11	381584		
12	379132	120000	p_2
13	376681		
14	374228		
15	371777		
16	369325		
17	379132		
18	376681		
19	374228		
20	371777		
21	369325	320000	p_3
22	366874		
23	364421		
24	361969		
25	359518		
26	357110		
27	366874		
28	364421		
29	361969		
30	359518		
31	357110	615000	p_4
32	354703		
33	352295		
34	349888		
35	347430		
36	345118		
37	342755		

MOMENTS.

$$M_x = Rx - m.$$

Span, 270 feet.

x	R	m	p
38	352295		
39	349888		
40	347480		
41	345118		
42	342755	109500	p_5
43	340391		
44	338029		
45	335665		
46	333347		
47	331028		
48	345118		
49	342755		
50	340391		
51	338029	190000	p_6
52	335665		
53	333347		
54	331028		
55	328711		
56	338029		
57	335665		
58	333347	250000	p_7
59	331028		
60	328711		
61	326392		
62	324074		
63	333347		
64	331028		
65	328711		
66	326392	327500	p_8
67	324074		
68	321800		
69	319525		
70	328711		
71	326392		
72	324074		
73	321800	407500	p_9
74	319525		
75	317251		
76	314977		

Span, 270 feet.

x	R	m	p
77	333347		
78	331028		
79	328711	565000	p_{10}
80	326392		
81	324074		
82	340391		
83	338029		
84	335665		
85	333347		
86	331028		
87	328711		
88	326392	717000	p_{11}
89	324074		
90	321800		
91	319525		
92	317251		
93	314977		
94	312703		
95	321800		
96	319525		
97	317251		
98	314977		
99	312703		
100	310428	824500	p_{12}
101	308155		
102	305880		
103	303607		
104	301332		
105	299102		
106	296873		
107	305880		
108	303607		
109	301332		
110	299102		
111	296873		
112	294643		
113	292414	944500	p_{13}
114	290184		
115	288000		
116	285814		
117	283629		

Span, 270 feet.

x	R	m	p
118	281444		
119	279258		
120	277118	944500	p_{13}
121	274977		
122	272836		
123	270695		
124	279258		
125	277118		
126	274977		
127	272836		
128	270695		
129	268555		
130	266458	1077000	p_{14}
131	264362		
132	262266		
133	260169		
134	258073		
135	255977		

For all values of x below x = 74, a uniformly distributed load of 2700 lbs. per ft. will give greater moments than this table.

Span, 280 feet.

x	R	m	p
1	401839		
2	399357		
3	396875		
4	394392	0	p_2
5	391910		
6	389471		
7	387032		
8	384592		
9	397100		
10	394606		
11	392156		
12	389706		
13	387257	120000	p_3
14	384806		
15	382357		
16	379950		
17	389706		
18	387257		
19	384806		
20	382357		
21	379950		
22	377542	320000	p_3
23	375135		
24	372727		
25	370321		
26	367956		
27	365593		
28	375135		
29	372727		
30	370321		
31	367956		
32	365593		
33	363228	645000	p_4
34	360863		
35	358500		
36	356135		
37	353771		
38	351406		

MOMENTS.

$$M_x = Rx - m.$$

SPAN, 280 feet.

x	R	m
39	360863	•
40	358500	
41	356135	
42	353771	
43	351406	1095000 f_2
44	348043	
45	346678	
46	344356	
47	342035	
48	339713	
49	337393	
50	351406	
51	348043	
52	346678	190000 f_6
53	344356	
54	342035	
55	339713	
56	337393	
57	346678	
58	344356	
59	342035	250000 f_7
60	339713	
61	337392	
62	335071	
63	344356	
64	342035	
65	339713	
66	337392	327500 f_8
67	335071	
68	332793	
69	330514	
70	339713	
71	337392	
72	335071	
73	332793	407500 f_9
74	330514	
75	328235	
76	325957	

SPAN, 280 feet.

x	R	m
77	344356	
78	342035	
79	339713	565000 f_{10}
80	337392	
81	335071	
82	351406	
83	348043	
84	346678	
85	344356	
86	342035	
87	339713	
88	337392	717000 f_{11}
89	335071	
90	332793	
91	330514	
92	328235	
93	325957	
94	323678	
95	332793	
96	330514	
97	328235	
98	325957	
99	323678	
100	321442	
101	319206	
102	316971	824000 f_{12}
103	314735	
104	312500	
105	310307	
106	308113	
107	305921	
108	303730	
109	301535	
110	310307	
111	308113	
112	305921	
113	303730	944000 f_{13}
114	301535	
115	299343	
116	297150	
117	294957	

SPAN, 280 feet.

x	R	m
118	292764	
119	290571	
120	288420	944500 f_{13}
121	286271	
122	284121	
123	292764	
124	290571	
125	288420	
126	286271	
127	284121	
128	281070	
129	279821	
130	277713	
131	275600	
132	273500	1077000 f_{14}
133	271392	
134	269286	
135	267220	
136	265156	
137	263092	
138	261028	
139	258963	
140	256943	

For all values of x below x = 61, a uniformly distributed load of 2700 lbs. per ft. will give greater moments than this table.

SPAN, 290 feet.

x	R	m
1	412113	
2	409675	
3	407237	
4	404800	0 f_1
5	402362	
6	399965	
7	397568	
8	395172	
9	407722	
10	405274	
11	402825	
12	400377	120000 f_2
13	397928	
14	395480	
15	393031	
16	390624	
17	400377	
18	397928	
19	395480	
20	393031	
21	390624	
22	388217	320000 f_3
23	385810	
24	383402	
25	380995	
26	378630	
27	376264	
28	385810	
29	383402	
30	380995	
31	378630	
32	376264	
33	373897	640000 f_4
34	371531	
35	369165	
36	366843	
37	364516	
38	362192	

MOMENTS.

$$M_x = Rx - m.$$

SPAN, 290 feet.			SPAN, 290 feet.			SPAN, 290 feet.			SPAN, 300 feet.		
x	R	m	x	R	m	x	R	m	x	R	m
39	371531		77	355260		118	303814		1	422463	
40	369165		78	352977		119	301655		2	420026	
41	366843		79	350694		120	299537		3	417599	
42	364516		80	348411		121	297420		4	415153	
43	362192	1093000 p_5	81	346128		122	295302		5	412716	
44	359868		82	343793	5650000 p_{10}	123	293185	944000 p_{13}	6	410320	0 p_0
45	357543					124	291067		7	407923	
46	355260					125	288950		8	405526	
47	352977		83	359868							
48	350694		84	357543							
49	348411		85	355260		126	297420		9	418080	
50	346128		86	352977		127	295302		10	415633	
			87	350694		128	293185		11	413226	
			88	348411		129	291067		12	410820	120000 p_2
51	359868		89	346128	7170000 p_{11}	130	288950		13	408413	
52	357543	1900000 p_6	90	343793		131	286832		14	406006	
53	355260		91	341509		132	284715	10770000 p_{14}	15	403600	
54	352977		92	339226		133	282597		16	401233	
55	350694		93	336943		134	280480				
56	348411					135	278402				
						136	276326		17	410820	
			94	346128		137	274250		18	408413	
57	357543		95	343793		138	272174		19	406006	
58	355260		96	341509					20	403600	
59	352977		97	339226					21	401233	
60	350694	2550000 p_7	98	336943		139	284715		22	398866	320000 p_3
61	348411		99	334660		140	282597		23	396500	
62	346128		100	332418	8240000 p_{12}	141	280480		24	394133	
63	343793		101	330177		142	278402	12800000 p_{15}	25	391766	
			102	327935		143	276326		26	389400	
			103	325693		144	274250		27	387033	
64	352977		104	323452		145	272174				
65	350694		105	321251							
66	348411		106	319051					28	396500	
67	346128	3271000 p_8	107	316851					29	394133	
68	343793		108	314651					30	391766	
69	341509		109	312451					31	389400	
									32	387033	
70	350694		110	321251					33	384666	64000 p_4
71	348411		111	319051					34	382300	
72	346128		112	316851					35	379933	
73	343793	4075000 p_9	113	314651	9445000 p_{13}				36	377666	
74	341509		114	312451					37	375280	
75	339226		115	310250					38	372953	
76	336943		116	308131							
			117	305972							

For all values of x below x = 31, a uniformly distributed load of 2700 lbs. per ft. will give greater moments than this table.

MOMENTS.

$$M_x = Rx - m.$$

SPAN, 300 feet.

x	R	m
39	382300	
40	379913	
41	377606	
42	375280	
43	372953	
44	370626	1095000 p_5
45	368300	
46	366013	
47	363726	
48	361440	
49	359153	
50	356866	
51	370626	
52	368300	
53	366013	1900000 p_6
54	363726	
55	361440	
56	359153	
57	368300	
58	366013	
59	363726	3350000 p_7
60	361440	
61	359153	
62	356866	
63	354620	
64	363726	
65	361440	
66	359153	327500 p_8
67	356866	
68	354620	
69	352373	
70	350126	
71	359153	
72	356866	
73	354620	407500 p_9
74	352373	
75	350126	
76	347880	
77	345633	

SPAN, 300 feet.

x	R	m
78	363726	
79	361440	
80	359153	565000 p_{10}
81	356866	
82	354620	
83	352373	
84	368300	
85	366013	
86	363726	
87	361440	
88	359153	
89	356866	
90	354620	7170000 p_{11}
91	352373	
92	350126	
93	347880	
94	345633	
95	343426	
96	341220	
97	350126	
98	347880	
99	345633	
100	343426	
101	341220	
102	339013	844000 p_{12}
103	336806	
104	334600	
105	332393	
106	330186	
107	327980	
108	325773	
109	334600	
110	332393	
111	330186	
112	327980	
113	325773	944000 p_{13}
114	323566	
115	321400	
116	319233	
117	317066	
118	314900	

SPAN, 300 feet.

x	R	m
119	312733	
120	310606	
121	308480	
122	306353	944900 p_{13}
123	304226	
124	302100	
125	300013	
126	308480	
127	306353	
128	304226	
129	302100	
130	300013	
131	297926	
132	295840	
133	293753	10770000 p_{14}
134	291666	
135	289620	
136	287573	
137	285526	
138	283480	
139	281433	
140	279386	
141	277340	
142	289620	
143	287573	
144	285526	1280000 p_{15}
145	283480	
146	281433	
147	279386	
148	277340	
149	275293	
150	283480	1432500 p_{16}

For all values of x below x = 45, a uniformly distributed load of 2700 lbs. per ft. will give greater moments than this table.

ILLUSTRATION OF THE USE OF THE TABLES.

A few illustrations will make clear the use of the tables.

1st, Single System of Bracing. — Take, for instance, the Pratt truss, worked out on p. 107, Fig. 99. Here we have $l = 90$ feet required to find the maximum strains due to live load.

For any upper flange, as A, Table II. for span 90 feet, p. xxii., gives at once the greatest moment. Thus, since the centre of moments for A is at the foot of the post ab, we must find the maximum moment at ten feet from the left end. For this point, the table gives at once,

$$\text{Maximum moment} = Rx - m = 157888 \times 10 - 125000 = 1453880;$$

dividing this by lever arm of A, which is 10 feet, we have at once,

$$A = 145388 \text{ pounds} = 72.69 \text{ tons.}$$

In like manner, for E, we have $x = 40$; and hence

$$E \times 10 = Rx - m = 145610 \times 40 - 2550000 = 3274400;$$

hence

$$E = 327440 \text{ pounds} = 163.72 \text{ tons.}$$

So for any flange, upper or lower, the greatest strain due to such a rolling load as our tables assume can be at once and easily found. This rolling load is believed to be such as to give greater strains than can ever occur in practice. To the strains thus found, we must, of course, add the strains due to dead load. If we assume this at 0.5 ton per foot, or 5 tons at each upper apex, as on p. 107, we have at once for A, since the re-action at left is 20 tons,

$$A \times 10 = 20 \times 10, \text{ or } A = 20 \text{ tons};$$

$$E \times 10 = 20 \times 40 - 5(30 + 20 + 10), \text{ or } E = 50 \text{ tons.}$$

The total strains are, therefore,

$$A = 72.69 + 20 = 92.69 \text{ tons};$$

$$E = 163.72 + 50 = 213.72 \text{ tons.}$$

Comparing these with the strains found on p. 107, viz., $A = 100.33$, and $E = 223$, tons, we see that the method given there gives us greater strains than our table. This would indicate that our "locomotive excess" of 33 tons, there adopted, is somewhat large, and that 28 tons would nearer meet the requirements of practice, and cause the values of p. 107 and those found by use of the table to agree quite closely.

For any diagonal, as fg, we can easily find the requisite shear from Table I. Thus, we have, in present case, $l = 90$, $x = 40$, $l - x = 50$, $l - x + 8 = 58$. We see at once, from

Table I., p. iii., that the limit for span 90 feet is 48⅓; hence, for $x = 40$, we must use formula II. We have, then,

$$\text{Shear} = \frac{15000}{90}58 + \frac{175000}{90}50 - \frac{3210000}{90} - 15000 = 56222 = 28.1 \text{ tons.}$$

To this add the dead-load shear, which is 5 tons, and we have shear for fg 33.1 tons. The shear found on p. 108, for 33 tons locomotive excess, is 40 tons. As before remarked, the results there given are somewhat large, and a locomotive excess of 28 tons would give a closer agreement.

The preceding, if followed carefully, will fully explain the use of our tables in all cases where we have but a single system of bracing.

2d, DOUBLE, OR MULTIPLE, SYSTEM. — For any flange, find from Table II. the greatest moment for its nearest centre of moments, and then find what uniform load will give this same moment at the same point ; calculate the flange for this uniform load divided into apex loads, and acting upon each system.

If x is the distance from left end to centre of moments, and p is the uniform load per unit of length, then $\frac{px}{2}(l - x)$ is the moment. Put this equal to the moment found from Table II., and the value of p can be found, in pounds per foot, which would give the same moment as the table, at the given point. The flange can then be calculated for this uniform load. Each flange should thus be calculated for its own equivalent uniform load.

For any brace, find the greatest shear at the foot of it from Table II. ; then find the equivalent uniform moving load ; divide this into apex weights, and calculate the brace for this loading.

If m is the uniform moving load per unit of length, coming on from right and extending up to a distance of x from the left end, then the shear is $\frac{m(l - x)^2}{2l}$.

Put this equal to the shear found from Table II., and the value of m can be found in pounds per foot, which would give the same shear as the table. The brace can then be calculated for this uniform moving load. Each brace should thus be calculated for its own equivalent uniform load.

www.ingramcontent.com/pod-product-compliance
Lightning Source LLC
Chambersburg PA
CBHW022016190326
41519CB00010B/1535